International Max Planck Research School (IMPRS)
for Maritime Affairs
at the University of Hamburg

Hamburg Studies on Maritime Affairs
Volume 5

Edited by

Ling Zhu

Compulsory Insurance and Compensation for Bunker Oil Pollution Damage

Springer

Dr. Ling Zhu
International Max Planck Research School (IMPRS)
for Maritime Affairs at the University of Hamburg
Max-Planck-Institut für ausländisches und internationales Privatrecht
Mittelweg 187
20148 Hamburg
Germany

ISSN 1614-2462

ISBN-10 3-540-45900-6 Springer Berlin Heidelberg New York
ISBN-13 978-3-540-45900-2 Springer Berlin Heidelberg New York

Springer is part of Springer Science+Business Media

springer.com

Production: LE-TeX Jelonek, Schmidt & Vöckler GbR, Leipzig
Cover-design: Erich Kirchner, Heidelberg

SPIN 11879565 64/3100YL - 5 4 3 2 1 0 Printed on acid-free paper

To my parents

Preface

On an international level, a distinction is made between the liability for tanker oil pollution damage and the liability for bunker oil pollution damage. Many books have been written on the former but not on the latter. The subject of my study may be deduced from the title without much difficulty. It is principally the study of bunker oil pollution damage with specific attention given to its compulsory insurance and compensation aspects. The basis of this study is the International Convention on Civil Liability for Bunker Oil Pollution Damage, 2001, which was adopted by a diplomatic conference at the International Maritime Organisation in March 2001. So far, it has not come into force. Eleven countries have ratified it and they are: Cyprus, Greece, Jamaica, Latvia, Luxembourg, Samoa, Singapore, Slovenia, Spain, The United Kingdom and Tonga. Among them, Singapore and the United Kingdom are the latest ratifying countries and not recorded in my study and thus deserve a special mentioning here. It is hoped that the findings of my study will provide readers with an interesting, clear and coherent picture of the Convention.

This study was conducted between October 2002 and April 2006 for a Ph.D. degree in law at the International Max Planck Research School (IMPRS) for Maritime Affairs and the University of Hamburg. I would like to firstly express my sincerest gratitude to my supervisor: Professor Dr. Dr. h.c. Jürgen Basedow, LL.M., Director of the Max Planck Institute for Foreign Private and Private International Law, Hamburg, for his faith in me to conduct the study, for his encouragement and for reading my drafts with incomparable patience, care and insight. Without his detailed comments during my research and writing, this final draft would not have been achieved. Grateful mention needs also to be made to my second examiner: Professor Dr. iur Gerrit Winter (now retired), for his valuable comments and the speedy submission of the second opinion on my thesis.

I am deeply indebted to the IMPRS for Maritime Affairs for accepting me as a scholar and sponsoring my study and stay in Germany very generously. I would also like to thank all IMPRS directors for their advice, support and encouragement as well as for accepting this thesis for publication in the IMPRS book series. Special thanks go to Professor Dr. Rainer Lagoni, LL.M., Director of the Law of the Sea and Maritime Law Institute of the University of Hamburg, and Professor Dr. Peter Ehlers, President of the German Federal Maritime and Hydrographic Agency, for their interesting seminars and related excursions to Helsinki and London.

Thanks are also due to my colleagues from IMPRS for Maritime Affairs: Benjamin, Deniz, Inken, Jennifer, Nicolai, Philipp, Rainer, Sicco, Suzette, Tilo, and many others who are not listed here. I wish to thank them all for their help, valuable suggestions and friendship in the course of the research leading to the completion of this study. I would also like to express my gratitude to our former and current coordinators – Dr. Wolfgang Wurmnest and Dr. Silke Knaut – for their tireless help. In particular, Dr. Knaut was of great help in accelerating the completion process of my study. Furthermore, I would like to acknowledge great help of Mrs. Gisela Schmitz-Krause, Sectary of Professor Jürgen Basedow, Mrs. Mary Pubusch, Librarian at the Law of the Sea and Maritime Law Institute, Mrs. Elke Hasen-Raffel, Librarian at the Max Planck Institute for Foreign Private and Private International Law and Mrs. Ellen Schaffer, Librarian at the International Tribunal for the Law of the Sea. I am additionally grateful to Mrs. Ingeborg Stahl for her precious help with editing the final version of this manuscript.

I wish to give my thanks and love to my parents, for their wholehearted belief, love and support, which motivated me to carry out my study. In addition, I am also grateful to my brother, my sister-in-law and my lovely little niece for their constant and enormous support for my study. Special thanks also go to my cousins, Dr. Chen, Yan Yun and Dr. Huang, Xiao Guang for their taking care of me during my LL.M. study at the University of Southampton and constant encouragement and help throughout my study.

There are many other friends to whom I wish to express my thanks. Their direct and indirect support helped me accomplish my thesis in time and enriched my stay both in Southampton and in Hamburg.

The effective date of this study's completion is April 2006. Unless otherwise indicated, the references were made before that time. I shall additionally remain responsible for any errors in the text.

Ling Zhu
Hamburg, August 2006

Contents

Abbreviations

A.C.	Appeal Cases
Act (1930)	The Third Parties (Right Against Insurers) Act 1930 (the United Kingdom)
Admin. L. J. Am. U.	Administrative Law Journal of American University
AMC	American Maritime Cases
Art.	Article
BIMCO	The Baltic and International Maritime Council
Bunkers Convention	International Convention on Civil Liability for Bunker Oil Pollution Damage
CERCLA	Comprehensive Environmental Response, Compensation and Liability Act of 1980 (the United States)
cf	compare
CLC	Civil Liability Convention
CLCs	1969 CLC and CLC Protocol 1992
1969 CLC	International Convention on Civil Liability for Oil Pollution Damage, 1969
CLC Protocol 1992	Protocol of 1992 to Amend the International Convention on Civil Liability for Oil Pollution Damage, 1969
CMI/C.M.I.	Comité Maritime International
COFR	Certificate of Financial Responsibility
Colum. L Rev.	Columbia Law Review
Convention	International Convention on Civil Liability for Bunker Oil Pollution Damage, 2001
1957 Convention	International Convention Relating to the Limitation of the Liability of Owners of Sea-Going Ships
CRISTAL	Contract Regarding a Supplement to Tanker Liability for Oil Pollution
Duke Envtl. L. & Pol'y F.	Duke environmental Law and Policy Forum
ed., eds.	editor, edition; editors, editions
e.g.	exempli gratia, for example
et.al.	et alii
etc.	et cetera
et seq.	et sequens
EU	European Union
F.N.(note)	footnote

Fund Conventions	The 1971 Fund Convention International Fund for Compensation for Oil Pollution Damage, 1971 and its 1992 Protocol
Geo. Int'l Envtl. L. Rev.	The Georgetown International Environmental Law Review
HL	House of Lords
HNS	International Convention on Liability and Compensation for Damage in Connection with the Carriage of Hazardous and Noxious Substances by Sea
ibid.	ibidem
ICC	Institute Cargo Clauses
IMCO	Inter-Governmental Maritime Consultative Organisation
IMO	International Maritime Organisation
IOPC(IOPCF)	International Oil Pollution Compensation
ITC	Institute Time Clauses
ITIA	International Tanker Indemnity Association
ITIC	International Transport Intermediaries Club
ITOPF	International Tanker Owners Pollution Federation Limited
IMO LEG	IMO Legal Committee
Int. I. L. R.	International Insurance Law Review
J.B.L.	The Journal of Business Law
J. Envtl.L.	Journal of Environmental Law
J. Mar. L. & Com.	Journal of Maritime Law & Commerce
K.B.	King's Bench (Law Reports)
La.L.Rev.	Louisiana Law Review
Law & Pol'y Int'L Bus.	Law and Policy in International Business
Ll.L.Rep.	Lloyd's List Reports (before 1951)
LLMC	Convention on Limitation of Liability for Maritime Claims
1976 LLMC	Convention on Limitation of Liability for Maritime Claims
LLMC Protocol 1996	Protocol of 1996 to Amend the Convention on Limitation of Liability for Maritime Claims, 1976
Lloyd's Mar. & Com. L.Q.	Lloyd's Maritime and Commercial Law Quarterly
Lloyd's Rep.	Lloyd's Law Reports
MARPOL	The International Convention for the Prevention of Pollution from Ships
Nat. Resources J.	Natural Resources Journal
OPA 90	Oil Pollution Act 1990 (the United States)
OPRC	International Convention on Oil Pollution Preparedness, Response and Co-operation
O.R.1969 CLC	Official Records of the International Legal Conference on Marine Pollution Damage, 1969
P&I	Protection and Indemnity

P&I Int.	P&I International
Sec.	Section
SDR	Special Drawing Rights
SOLAS	International Convention for the Safety of Life at Sea
STOPIA	Small Tanker Oil Pollution Indemnification Agreement
TOVALOP	Tanker Owners' Voluntary Agreement Concerning Liability for Oil Pollution
Tul. Mar. L. J	Tulane Maritime law Journal
Tul. L. Rev.	Tulane Law Review
UNCLOS	United Nations Convention on the Law of the Sea 1982
UK/U.K.	United Kingdom
US(A)	United States (of America)
U.S.C	United States Code
viz.	videlicit
vol.	volume
W.L.R.	Weekly Law Reports
WTO	World Trade Organisation
WWF	World Wildlife Fund for Nature

Introduction

Maritime transportation is undoubtedly a service of utmost significance. Its importance is primarily anchored in its being an indispensable service involving different States. With the aim of regulating and facilitating this service, there is a need to establish a uniform set of rules and regulations in many respects in relation to maritime transportation.

Maritime transportation has an inevitable impact on the marine environment. One of the most distinct impacts is pollution which might be caused by the spill or discharge of oil from ships. It is generally recognised that preventing pollution is better than compensating for the resulting damage. Specific standards and regulations have thus been set in place to control or prevent oil pollution from ships. One example is the International Convention for the Prevention of Pollution from Ships, 1973, as modified by the Protocol of 1978 relating thereto (MARPOL 73/78).[1] It is one of the most important international conventions for the prevention of pollution of the marine environment caused by ships. Annex I of the MARPOL 73/78 specifically provides regulations for the prevention of pollution by oil.[2] However, despite those well-established regulations as well as other developed standards, marine pollution incidents continue to occur. None of these regulations provide civil liability and compensation rules which was thus the purpose of international civil liability conventions.

Among the various types of ships, the tanker is the target of a set of strict liability rules under international civil liability conventions.[3] It subjects all the

---

1 According to the International Maritime Organization (IMO): as the 1973 MARPOL Convention had not yet entered into force, the 1978 MRPOL Protocol absorbed the earlier convention. The combined instrument – the International Convention for the Prevention of Maritime Pollution from Ships, 1973 as modified by the Protocol of 1978 relating thereto (MARPOL 73/78) – finally entered into force on 2 October 1983 (for Annexes I and II). The information is available at: <http://www.imo.org/Conventions/contents.asp?doc_id=678&topic_id=258#2> (visited 10 January 2006). They are reprinted in: *Lloyd's Shipping Law Library: The Ratification of Maritime Conventions* (2004), Vol.4, II.7.160 and II.7.170.

2 The Annex I entered into force on 2 October 1983 and the revised Annex I will enter into force on 1 January 2007.

3 The international civil liability convention system includes: the International Convention on Civil Liability for Oil Pollution Damage 1969 and its 1992 Protocol, and the International Convention on the Establishment of an International Fund for Compensation for Oil Pollution Damage 1971 and its 1992 Protocol. They are reprinted in: *Lloyd's Shipping Law Library: The Ratification of Maritime Conventions (2004)*, Vol.4, II.7.30, II.7.51, II.7.90 and II.7.111.

persons involved to a uniform set of legal rules. Strict liability is imposed on the owner of the oil tanker. In addition, the tanker owner is required to take out insurance to cover his liability for potential oil pollution risks. No such obligations are imposed on the owners of any other types of ship. The uniform liability rule for other types of ship, however, would have a similar effect on the persons involved in being liable for oil pollution damage from ships' bunkers. There, thus, remains a gap in the otherwise comprehensive international regime of liability and compensation for oil pollution from ships. Due to the absence of uniform rules, the relevant problems such as the nature of liability, the commensurate amount of compensation for victims, the nature and extent of the insurance risk, etc. remain uncertain. Solutions in this regard may promote the smooth operation of maritime transportation. National legislation might exist in some countries, but it is not uniform. Most countries do not have any specific legislation to deal with liability and compensation issues in relation to bunker oil spills. Therefore, uniformity is necessary and important.

In order to fill in the last above-mentioned gap, the International Convention on Civil Liability for Bunker Oil Pollution Damage (the Bunkers Convention) was adopted following a diplomatic conference at the International Maritime Organization (IMO) in March 2001. However, it has not yet come into force. The Bunkers Convention follows the format of earlier international civil liability conventions. It establishes a liability, insurance and compensation regime for spills of oil carried as fuel in ships' bunkers.

The insurance covering oil pollution risk arising from the operation of ships falls within the ambit of marine insurance. This type of insurance undertakes to indemnify the assured against the losses incident to marine adventure. However, since international civil liability conventions are adopted to ensure the payment of effective and adequate compensation to victims, the compulsory insurance requirements under the Bunkers Convention should also take into account the benefits of pollution victims. A direct and enforceable right to claim against the insurer is thus conferred on eligible claimants.

The insurance industry introduces or improves policies and integrates established insurance practice in order to participate as the insurer of the new risks. In particular, the P&I Club,[4] the main liability insurer of the shipowner, offers insurance coverage which reflects the shipowners' liabilities and it has over the years displayed flexibility, since it permits the inclusion of new risks such as pollution risk, P&I insurance likewise includes oil pollution risks. The practice of insuring tanker oil pollution liability shows the role and viability of P&I Clubs. They pay compensation in the case of pollution damage resulting from oil pollution incidents.

Under the insurance arrangement in accordance with the Bunkers Convention, three main parties, namely, the liability insurer, the shipowner and pollution

[4] P&I Clubs are shipowners' Protection and Indemnity Clubs, which will be given more detailed discussion in chapter 2 and appear quite often in the following chapters.

victims, will be involved.[5] The harmonisation and balance of their interests will be very important, as it contributes to establishing and maintaining an effective liability and compensation system.

A. The aim of the research

Insurance against oil pollution from ships has a compulsory nature under the Bunkers Convention. Except for the fact that the obligation to take out insurance is imposed on the registered owner, the Bunkers Convention also provides that the liability insurer will pay, on behalf of the assured, to pollution victims all sums equal to the applicable limit of liability.[6] In this way, the victims are assured of the availability of the compensation fund.

Against this background, the aim of this research is to complete a study on compulsory insurance and compensation for bunker oil pollution damage under the Bunkers Convention. Based on a general survey of the Bunkers Convention mainly focusing on the insurance and compensation aspects, this thesis will explore in-depth the role of liability insurance as a means to ensure compensation for victims who will suffer bunker-oil pollution damage. Since the main features of the Bunkers Convention, like in other civil liability conventions, are the compulsory insurance, strict liability and the limitation of liability, the thesis will also examine the interrelations between compulsory insurance and those other features.

B. The scope of the research

Due to the massive pollution damage that may result from oil pollution incidents, international rules as regards civil liability and compensation for oil pollution from oil tankers have been set up. The introductory chapter of this thesis is a description of major developments and the features of the international civil liability conventions system. On the basis of the brief description of earlier industry solutions and conventions, a detailed overview of the Bunkers Convention will be given in the same chapter.

The revolutionary idea of compulsory insurance for oil pollution came into being with the establishment of the International Convention on Civil Liability for Oil Pollution Damage, 1969 (1969 CLC). However, the requirement to take out compulsory insurance or other financial security does not automatically create a market for it. Therefore, the availability of the insurer is crucial for the efficiency

5 The interests of victims, the shipowner and his insurer are more delicate and direct than other possible interests under the international civil liability convention system such as the interests of different States.

6 The Bunkers Convention, Art.7(1): "...but in all cases, not exceeding an amount calculated in accordance with the Convention on Limitation of Liability for Maritime Claims, 1976, as amended."

of the operation of liability insurance and compensation regime in the Bunkers Convention. Chapter 2 begins with the examination of the structure of the marine insurance market in general. Since insurance coverage limits of other types of marine insurance become manifest, the importance of shipowners' P&I Clubs to cover shipowners' liability incurred in relation to the operation of ships is highlighted. The operation of P&I Clubs and their role in relation to oil pollution will likewise be illustrated in the said chapter.

The operation of compulsory insurance should comply with the general principles of marine insurance law. Its operation will have to be confined within the framework of the Bunkers Convention as well. Chapters 3, 4, 5 and 6 will be devoted to the examination of the relations between insurance and the strict liability principle, insurance and compensation, insurance and limitation of liability as well as the right of direct action.

Two questions are foremost in this research: 1) who should be liable? and 2) who is in the most convenient position to take out insurance? Under other international civil liability conventions, the party who can most easily be identified is regarded as the party who is most suitable to be liable and take out insurance. For the same reason, the registered owner of the ship shall be strictly liable and shall take out insurance. However, other parties such as the bareboat charterer, manager and operator shall also be strictly liable for pollution damage under the Bunkers Convention.[7] Chapter 3 will give a more detailed analysis of the issue of the strict liability rule, and examine whether it is compatible with the insurance arrangement.

In order to guarantee the availability of compensation to the greatest extent, the requirement of compulsory insurance is imposed on a great number of ships of different types. However, P&I Clubs have noticed the enormous bureaucracy required to administer a compulsory insurance scheme. This is largely due to the fact that the Bunkers Convention will, once ratified, involve different types of ships having a gross tonnage greater than 1,000 in a State Party. There is, thus, a conflict of interests between the benefits to third parties and the extra costs of administration of the Clubs and the State authorities involved in issuing compulsory insurance certificates. Chapter 4, therefore, will try to explore the possibility of other alternatives to the required insurance certificate. Due to the uncertainty of the adequacy of the compensation, other kinds of arrangement for compensation to pollution victims will also be suggested in this chapter.

In the absence of a fund similar to the International Oil Pollution Compensation Fund (IOPC Fund) under the Bunkers Convention,[8] ratifying States of the Bunkers Convention may require further reassurance that the recovery will always be possible and adequate. However, the insurer is reluctant to provide insurance coverage for oil pollution damage unless a well-defined upper limit on losses is established. Accordingly, Chapter 5 will focus on the analysis of the interrelation

[7] The Bunkers Convention, Art.1(3) and Art.3(1).

[8] The IOPC Fund was created in order to provide supplementary compensation to the pollution victims in s tanker oil spill incident. It will be considered in more detail in Chapter 1, Section A.II.

between limitation of liability and the limit of insurance. Chapter 6, on the other hand, will focus on the examination of a fundamental right of pollution victims to claim directly against the insurer.

Finally, the last Chapter will conclude the thesis by considering that it is important to strike a balance of interests among the pollution victims, the shipowner and the liability insurer under the oil pollution insurance and compensation system.

C. Methods used for the research

Since the purpose of this thesis, as the title indicates, is to examine compulsory insurance and compensation for bunker oil pollution damage caused by ships other than oil tankers, the thesis will mainly focus on relevant aspects embodied in the Bunkers Convention. However, the survey of relevant aspects will not be limited to the text of the Bunkers Convention itself. Reference to legal documents from the Legal Committee of the IMO will also be made.

An exploration of the laws relating to liability for bunker oil pollution damage cannot be accomplished without alluding to the earlier international civil liability conventions, which served as a model for the drafting of the Bunkers Convention. The relevant provisions of the earlier conventions will thus frequently be used as references while discussing the Bunkers Convention.[9] In addition, only little case law exists as far as the bunker-oil spill incident from non-tanker ships is concerned. Hence, only a handful of cases will appear in this research, a number of which are borrowed from tanker oil-pollution cases and a few from the legal documents of the IMO.

The law of and practice in marine insurance have a great impact on the earlier international civil liability conventions, as well as on the Bunkers Convention. Since the liability insurance coverage for marine risks is dominated by P&I Clubs, the examination of the Club and its role in insuring against oil pollution risk is thus necessary. This examination will be based on the whole background of the marine insurance market in general and will largely focus on the Club Rules and their practice. The detailed analysis of insurance and its relevant capacity is based on the current practice of tanker oil pollution. The literature in this respect is mainly from the London marine insurance market, since it is the largest insurance market in the world. However, references will also be made to the laws of other countries when dealing with some specific issues.

No book has yet been published concerning insurance and compensation for bunker-oil pollution. Thus, this research work is timely and necessary. Since the Bunkers Convention has not yet come into force, this research work shall not only focus on the analysis of the basic principles and issues relating to compulsory insurance and compensation as provided for in the Bunkers Convention, but also

---

9 However, it is important to point out that the provisions as regards some aspects in the Bunkers Convention are different.

try to explore the reasons for its current low ratification, and thus may show a possible way of improving the situation.

Chapter 1: Pollution from Ships' Bunkers and the Advent of the Bunkers Convention

A. A brief history of the development of the oil spill civil liability system

The year 2001 was a milestone in the development of the international system of liability and compensation for oil-pollution damage, due to the Bunkers Convention being adopted at a Diplomatic Conference at the IMO on 23 March 2001. The aim of the Convention is to ensure the availability of adequate, prompt and effective compensation to persons who suffer damage caused by oil spills when the oil is carried as fuel in ships' bunkers. By covering all non-tanker vessels, the Convention fills in the last significant gap left in an otherwise well-established international system of liability and compensation for oil pollution caused by ships.

Despite the best-developed preventive measures, marine pollution incidents continue to occur. As far as tanker oil-pollution incidents are concerned, there are numerous incidents involving relatively minor spills, but there are occasionally some catastrophic incidents resulting in severe pollution damage to the marine environment. The origin of the current international legal regime for liability and compensation for oil pollution was due to the *Torrey Canyon* disaster. The *Torrey Canyon*, a Liberian-registered tanker, went aground on the Seven Stones reef between the Scilly Isles and Land's End on 18 March 1967. As a result of the incident, a considerable amount[1] of crude oil cargo escaped and caused extensive damage. The spill severely affected nearby beaches, seabirds and the aquatic environment. The disaster consequently turned out to be a decisive factor in the

---

1 See Documentation C.M.I. 1968 – I, at 68: "2 The stranding damaged many of the cargo tanks and by March 20 it was estimated that 30,000 tons of oil had spilled into the sea. On March 25, oil began to arrive on the Cornish beaches, 100 miles of coastline being affected. On March 26 high seas and strong winds caused the ship to break her back, releasing, it is estimated, a further 30,000 tons of crude oil. Between March 28 and 30 the ship…" It was likely that more oil escaped afterwards. For instance, in: Braekhus, Sjur/Rein, Alexander, *Handbook of P&I Insurance* (1979), p. 184, it is said that 120,000 tons of crude oil escaped and caused extensive damage. As well, on the IMO website, *Torrey Canyon* is said to have spilled her entire cargo of 120,000 tons of crude oil into the sea: see <http://www.imo.org/ Environment/mainframe.asp?topic_id=231> (visited 19 May 2005).

development of two voluntary agreements and two international civil liability conventions. They ensured adequate compensation to persons who suffered damage caused by pollution resulting from the escape or discharge of persistent hydrocarbon mineral oil from ships. This aspect has already been covered in a vast amount of literature and this research confines itself to a general description of the development and major features of those compensation schemes.

I. TOVALOP and CRISTAL

Prior to the uniform set of international conventions, two voluntary agreements were established to play an initial constructive role in the compensation for oil-pollution damage from tankers. The voluntary agreements were known respectively as TOVALOP (Tanker Owners' Voluntary Agreement Concerning Liability for Oil Pollution) and CRISTAL (Contract Regarding a Supplement to Tanker Liability for Oil Pollution).[2] The agreements were established by the tanker and oil industries in the late 1960s in response to the problems highlighted by the *Torrey Canyon* incident. They were established as interim arrangements pending the ratification of international conventions.

These agreements bound the parties concerned to compensate oil pollution victims, regardless of fault, up to certain limits. They also provided that the parties to the agreements should obtain insurance for oil-pollution liabilities. The insurance cover was arranged through the shipowners' P&I Clubs or ITIA (International Tanker Indemnity Association Limited). The latter was set up to arrange insurance which enables a TOVALOP member to meet his obligations under the terms of the agreement.[3]

After coming into effect, both agreements played quite an important role in countries where international civil liability conventions were not in force. Despite the subsequent adoption of the CLCs and the Fund Conventions,[4] TOVALOP and CRISTAL still remained in force and functioned as a parallel system to the later

---

2 The TOVALOP and the CRISTAL are reprinted in: *Lloyd's Shipping Law Library: The Ratification of Maritime Conventions* (2004), Vol. 4, II.7.60 and II.7.120. The TOVALOP came into effect in October 1969. Following the establishment of the TOVALOP, oil-cargo interests established the CRISTAL in 1971 to provide additional compensation over and above that available under the TOVALOP where the cargo was owned by a party to the CRISTAL.

3 ITIA is a mutual P&I insurer set up in the aftermath of the *Torrey Canyon* disaster of 1976. It provided cover for liabilities assumed under TOVALOP and those for which the owner was legally liable under statute or otherwise by reason of a discharge or threatened discharge of oil. See Wu, Chao, *Pollution from the Carriage of Oil by Sea: Liability and Compensation* (1996), p. 110.

4 The CLCs refer to the International Convention on Civil Liability for Oil Pollution Damage, 1969 and its 1992 Protocol. The Fund Conventions denote the International Convention on the Establishment of an International Fund for Compensation for Oil Pollution Damage, 1971 as well as its 1992 Protocol. More see Section A.II of this chapter.

conventions. Moreover, as the CLCs and the Fund Conventions became widespread, both agreements readjusted the claim settlements between shipowners and oil companies in order to bring them in line with international conventions.

With the evolution of voluntary agreements themselves,[5] the compensation level was changed. In the original version of TOVALOP, there was a limit of US$100 per gross ton or US$10 million in total, whichever was less. The P&I Clubs agreed to provide such cover.[6] Under the TOVALOP Standing Agreement, the maximum compensation for all claims arising out of any one incident was raised to US$160 per limitation ton or US$16.8 million, whichever was less.[7] In 1987, the new TOVALOP Supplement was created. It was not for the purpose of replacing the original TOVALOP. The TOVALOP Supplement applied only to incidents when a participating tanker was carrying cargo owned by a party of CRISTAL.[8] After February 20, 1994, the limits of financial responsibility under the TOVALOP Supplement were denominated in Special Drawing Rights (SDR), the value of which was calculated in accordance with the method of valuation used by the International Monetary Fund.[9] The maximum amount to be paid by the shipowner for each incident was not to exceed SDR 3 million for a ship of 5,000 gross tons or less; for ships in excess of 5,000 gross tons, SDR 420 was to be paid for each additional ton over and above the SDR 3 million, but should in no case exceed SDR 59.7 million, which was the maximum amount to be paid by the shipowner.[10]

The CRISTAL only intervened in a limited number of circumstances. Before 1987, the total maximum amount that the CRISTAL fund could pay out as compensation per incident was US $36 million. This figure included the total amount of compensation paid by the shipowner and any other liable party.[11] In the 1994 revision, the CRISTAL limits were summarised as follows: SDR 32 million for a ship of 5,000 gross tons or less; and an additional SDR 652 for every gross ton in excess, but in no case was it to exceed SDR 120 million, which was the

5 Wu, Chao, *supra*, note 3, p. 105: "The attempt to bring the voluntary regime into line with the convention system was to ensure that the same universal level of liability applied to all shipowners worldwide." See also Abcassis, David W. (ed.), *Oil Pollution from Ships* (1985), p. 305: "... and other changes were made so that the agreement mirrored the Liability Convention very closely." The changes were made so that the agreements could be brought into line with the convention system and ensure the same universal level of liability applied to all shipowners worldwide.

6 Wu, Chao, *ibid.*, or Abecassis, David W.(ed.), *ibid.*

7 TOVALOP Standing Agreement, reprinted in: *Lloyd's Shipping Law Library*: *The Ratification of Maritime Conventions* (2004), Vol. 4, II.7.60, Art. VII(A).

8 TOVALOP Supplement, Clause 1(1)(A): "Applicable Incident", reprinted in *ibid.*

9 The SDR is an international reserved asset, created by the International Monetary Fund in 1969. Its value is based on a basket of key international currencies. More information is available at: <http://www.imf.org>.

10 *Ibid.*, Clause 3(c)(3).

11 Wu, Chao, *supra*, note 3, p. 119.

maximum amount that the CRISTAL fund was to pay. The figure included the amount of compensation paid under TOVALOP.

Although the importance of the voluntary agreements was progressively eroded by the widespread ratification of equivalent conventions around the world, they lasted far longer than anyone expected. The voluntary agreements of TOVALOP and CRISTAL were ultimately terminated on 20 February 1997, after having been in force for 28 and 26 years respectively.

II. International Conventions

The *Torrey Canyon* disaster exposed the deficiencies in the legal regime providing compensation following an oil-pollution incident at sea. At that time, there was no international convention dealing with liability and compensation for oil pollution from ships. Immediately after the incident, both the British and French Governments raised the matter of marine pollution with the Inter-governmental Maritime Consultative Organization (IMCO). As a consequence of the said incident, the IMCO set up its own legal committee to deal with oil pollution.[12] In November 1969, the International Convention on Civil Liability for Oil Pollution Damage emerged at the Brussels Conference.[13] Six years later, the International Convention on Civil Liability for Oil Pollution Damage, 1969 (1969 CLC) came into force and provided a uniform international regime to compensate those who suffer pollution damage as a result of the escape or discharge of persistent oil from laden tankers. In 1978, the International Convention on the Establishment of an International Fund for Compensation for Oil Pollution Damage, 1971 (1971 Fund Convention) came into force[14] in order to provide a supplementary international regime, financed by the receivers of crude and heavy fuel oil carried by sea.

Thus, from 1978 to 1997, there were two international legal regimes (the Liability and Fund Conventions) and two voluntary agreements (TOVALOP and CRISTAL) through which compensation for pollution damage was available following spills of persistent oil from oil tankers.

After several disasters,[15] it was acknowledged that the limits of compensation available under international conventions were insufficient to meet all reasonable claims in the event of a substantial oil spill incident. Accordingly, in May 1984, the Protocols to the Liability and Fund Conventions were agreed upon. They substantially increased the limits of compensation and coverage from the earlier 1969 CLC and the 1971 Fund Convention. These Protocols, however, never came into force. In November 1992, the original Conventions (the 1969 CLC and the

---

12 *Ibid.*, p. 37.

13 The Brussels Conference was held on 10-28 November 1969. The objective of the Conference was to examine the drafts of the international conventions proposed to deal respectively with the public law and private law issues highlighted by the *Torrey Canyon*.

14 This Convention ceased to be in force on 24 May 2002.

15 For example: *Amoco Cadiz* on 18 March 1978, *Tanio* on 7 March 1980.

1971 Fund Convention) were revised to facilitate ratification.[16] The new Protocols have since then been known as the 1992 Protocols.

The present dominant international conventions are the 1969 CLC as amended by the 1992 Protocol thereto and the 1971 Fund Convention as amended by the 1992 Protocol thereto. As of 31 December 2005, 113 States are Contracting Parties to the CLC Protocol 1992 and 98 States are Contracting Parties to the Fund Protocol 1992.[17]

The fundamental features of international civil liability conventions are the establishment of strict liability, limitation of liability and compulsory insurance. Under the CLCs, the victims in a tanker oil-pollution incident could easily identify the liable person since they impose the liability for oil-pollution damage squarely on the registered owner[18] of the ship from which the oil had escaped or was discharged. This method is known as the "channelling mechanism" accompanying the shipowner's right of recourse, which will be discussed later.[19] The liability of the shipowner is strict in the sense that he is liable irrespective of the existence of any fault. In other words, the claimant has only to demonstrate that he has suffered pollution damage as a result of the spill; there is no need to prove the shipowner's negligence. Ships carrying more than 2,000 tons of persistent oil as cargo are required to maintain appropriate liability insurance or other financial security. A right of direct action is also established for the claimants against the insurer or other guarantors. All these measures benefit the victims. They facilitate prompt and equitable compensation payments to the victims. In return, the shipowner is strictly liable only under the CLCs and liability outside the CLCs is excluded. The shipowner may limit his liability, although this right may be lost in some extreme circumstances.[20]

---

16 DelaRue, Colin/Anderson, Charles B., *Shipping and the Environment* (1998), p. 71, it is of an opinion that the ratification was also quickened by two major incidents occurring within the weeks of the conference. These two incidents were: *Aegean Sea* on 3 December 1992 and *Braer* on January 1993.

17 The information is available at: <http://www.imo.org/Conventions/mainframe.asp?topic_id=247> (visited 3 January 2005).

18 The "owner" is defined in the 1969 CLC, Art. I(3): "…means the person or persons registered as the owner of the ship or, in the absence of registration, the person or persons owning the ship. However, in the case of a ship owned by a State and operated by a company which in that State is registered as the ship's operator, 'owner' shall mean such company." Meanwhile, "person" in Article I(2) means "any individual or partnership or any public or private body, whether corporate or not, including a State or any of its constituent subdivisions." There was no change in the CLC Protocol 1992.

19 Shipowner's right of recourse is provided by Art. III (5) of the 1969 CLC, which provides: "Nothing in this Convention shall prejudice any right of recourse of the owner against third parties." There is no change in the CLC Protocol 1992.

20 Under the 1969 CLC, the shipowner was able to limit his liability only if he could prove that the incident had not resulted from his "actual fault or privity". In the CLC Protocol 1992, the owner shall not be entitled to limit his liability if it is proved that the pollution damage resulted from his personal act or omission committed with the intent to cause such damage, or recklessly and with knowledge

It was realised that the consequence of oil pollution damage should not exclusively be borne by the shipping industry. Part of the liability arising from an oil spill incident should be shared by oil-cargo interests. In order to accompany the CLCs and in effect balance conflicting interests, a second tier of compensation is provided under the Fund Conventions. The purpose of the Fund Conventions is to create a fund to provide a supplementary compensation, available in cases where the totality of claims exceeds the shipowner's liability limit or where the compensation cannot be obtained from the shipowner, for instance, if the shipowner is exonerated from liability. For this reason, the IOPC Fund was created under the Fund Convention.[21] The said fund is financed by contributions of oil companies or other entities that receive crude oil and heavy fuel oil after sea transport. Accordingly, the CLCs are not solely responsible for the liability and compensation for oil pollution damage. There used to be two IOPC Funds: the 1971 Fund and the 1992 Fund. They co-existed together for quite a while, and since the 1971 Fund Convention, as well as its Fund, ceased to be in force on 24 May 2002, the role of the 1971 Fund was weakened and terminated with its complete replacement by the 1992 Fund.

All in all, those affected by spills of persistent oil from tankers benefit from a uniquely successful international two-tier system of compensation, as illustrated above. The liability of tanker owners under the CLC Protocol 1992 ranges from SDR 3 million for a small tanker up to 5,000 gross tons to SDR 59.7 million for a ship of 140,000 units of tonnage or over. A maximum of SDR 135 million is also available for victims per incident from the 1992 Fund, irrespective of the size of the tanker. The figures include the sum paid by the tanker owner or his insurer under the 1992 Protocol.[22]

---

that such damage would probably result. It is obvious that the CLC Protocol 1992 adopted a more restrictive method of barring the shipowner's right of limitation.

21 The 1971 Fund Convention, Art.2: "An International Fund for compensation for pollution damage, to be named, 'The International Oil Pollution Compensation Fund' and hereinafter referred to as 'The Fund', is hereby established with the following aims:..."

22 Three points are worthy of notice:
(1) In October 2000, the IMO Legal Committee agreed to increase compensation limits provided by the 1992 CLC/Fund Convention by approximately 50 percent. This increase came into force on 1 November 2003.
(2) In May 2003, it was agreed at a diplomatic conference to adopt *a third tier of compensation* to supplement the 1992 CLC/Fund regime to be called the "2003 Protocol on the Establishment of a Supplementary Fund for Oil Pollution Damage". The third tier will provide compensation up to SDR 750 million, including the compensation provided under the 1992 CLC/Fund compensation scheme. In addition, the third tier is a voluntary level of compensation which may be ratified by any State Party to the 1992 CLC/Fund compensation scheme, and was to come into force three months after at least eight countries representing no less than 450 million tons of contributing oil imports ratify it. It entered into force on 5 March 2005.

B. The need for the Bunkers Convention

This section will discuss, in four aspects, the legislative history which triggered the adoption of the Bunkers Convention. They are: (1) the scope of earlier conventions; (2) the background work of the Bunkers Convention; (3) the technical considerations; and (4) the advent of the Bunkers Convention.

I. The scope of earlier conventions

The main purpose of the Bunkers Convention is to establish a liability and compensation regime for spills of oil carried as fuel in ships' bunkers. The question, however, is whether pollution damage caused by bunker-oil spills has already been covered by earlier conventions. The international conventions discussed in earlier sections show that the established regime covering civil liability and compensation for oil spills does not include those from vessels other than tankers. The reasons for such a conclusion are:

Article 1(1) of the 1969 CLC defined "ship" as being "any sea-going vessel and any seaborn craft of any type whatsoever, actually carrying oil in bulk as cargo"[23] The said definition was replaced by the CLC Protocol 1992 thereto with the following text:

> "'Ship' means any sea-going vessel and seaborne craft of any type whatsoever constructed or adapted for the carriage of oil in bulk as cargo, provided that a ship capable of carrying oil and other cargoes shall be regarded as a ship only when it is actually carrying oil in bulk as cargo and during any voyage following such carriage unless it is proved that it has no residues of such carriage of oil in bulk aboard."[24]

Accordingly, depending on whether the 1969 CLC or the CLC Protocol 1992 is in force, the liability for a bunker-oil spill may be covered by the CLCs in some limited situations. A bunker spill from a laden tanker would be covered by the 1969 CLC. The CLC Protocol 1992 enlarged the scope of application; it applies to

---

(3) To maintain the balance of sharing liabilities between the tanker and oil industries achieved by the 1992 CLC/Fund, the International Group of P&I Clubs has also agreed to a voluntary increase in the minimum limit of liability applicable under the CLC Protocol 1992 to *small tankers* (currently SDR 3 million as discussed above) to SDR 20 million. This so-called "Small Tanker Oil Pollution Indemnification Agreement" (STOPIA) came into effect simultaneously, and in the same countries, as the third tier of compensation discussed above. The indemnity under the STOPIA will, however, only apply in the event of tanker spills affecting a State in which the Supplementary Fund Protocol is in force and when liability is imposed on the shipowner under the CLC Protocol 1992. Neither the flag of the vessel nor the ownership of the cargo is relevant.

23 The 1969 CLC, Art. I(1), the same definition is adopted by the Fund Convention 1971, see Art.1.2.

24 The CLC Protocol 1992, Art.2(1), the same definition was adopted by the Protocol of 1992 to amend the Fund Convention.

spills of bunker oil from unladen tankers in certain circumstances. However, as seen from the definition of "ship", the CLC Protocol 1992 is only intended to apply to bunker spills from such a vessel when it is sailing in ballast while operating in the oil trade, but not when it is engaged in the carriage of other types of goods. Therefore, neither the 1969/1971 conventions nor the 1992 protocols apply to spills of bunker oil from ships other than oil tankers.[25] This principle even applies to the case in which the origin of the pollution is not clear. In this case, according to the IOPC Fund, neither the CLCs nor the Fund Conventions are applied. On 28 and 29 September, 1997, when bunker fuel oil reached the sandy beaches of Essex on the east coast of England, the claim for compensation for the cost of the clear-up operations was submitted to the 1992 IOPC Fund, provisionally indicated at approximately €10,000. Since the origin of the oil was unknown, and in view of the fact that only a small quantity of oil had landed on the beach, the director of the IOPC Fund considered it was unlikely to establish whether the oil came from a tanker, whether it was laden or unladen. For this reason, the claim was not pursued.[26]

The cases involving combination carriers also exemplify some defects in the current civil liability convention system, since they may operate as tankers or as non-tankers. The combination carriers are covered by either version of the CLCs when they are trading as tankers. However, if they are not carrying any oil cargo, and pollution is caused by the escape of bunker oil, the 1969 CLC will not apply to them, but the CLC Protocol 1992 may cover the resulting pollution damage in some cases, since the "voyage following such carriage" in the definition of "ship" in the CLC Protocol 1992 usually means a ballast passage. However, in the case of combination carriers, it might be the case where oil cargo is discharged and a dry cargo is then loaded at the same port – with the result that the vessel's next voyage is in fact in a dry cargo trade. Therefore, if "it is proved that it has no residues of such carriage of oil in bulk aboard", the CLC Protocol 1992 will not apply to any bunker spill from the ship during that voyage in this case.[27]

II. National legislation and the background work on the Bunkers Convention

The purpose of the Bunkers Convention, like any other international civil liability conventions, is to provide a set of rules that will be applied uniformly in all contracting States to the Convention. As the title of this section shows, the status of national legislation before the adoption of the Bunkers Convention will be

---

25 In the 1992 Fund Convention, Art. 2(3): "Ship", "Person", "Owner", "Oil", "Pollution Damage", "Preventive Measures", "Incident", and "Organization" have the same meaning as in Article 1 of the 1992 Liability Convention.

26 See IOPC Fund Annual Report 1998, at 111; IOPC Fund Annual Report 1997, at 126.

27 DelaRue Colin M./Anderson, Charles B. *supra*, note 16, p. 80.

briefly reviewed, and the background work on the Bunkers Convention on an international level will also be discussed.

On a national level, one may find that legislation for handling bunker-oil pollution liability is in force in some coastal States. For instance, the United Kingdom, in order to rectify the inequity between the liability imposed on tankers and non-tankers, has extended the CLC regime to bunker spills.[28] The United States adopted the Oil Pollution Act of 1990 (OPA 90), which deals with oil pollution from all types of vessels.[29] However, many countries do not have domestic legislation to deal with liability and compensation for bunker-oil pollution damage. If the legislations in this respect are not uniform at the international level, the involved parties may be unclear as to the extent of their rights and liabilities. For instance, shipowners would be exposed to divergent domestic laws. Thus, they might decide to flee to the jurisdiction with less demanding legislation. Another consequence might be that the pollution victims will face difficulties in finding the liable person and in choosing the jurisdiction to file the claim and so on.

On an international level, there was no uniform legislation until the adoption of the Bunkers Convention. As a matter of fact, pollution damage caused by bunker-oil spills started to be taken into account following the preparatory work of the 1969 CLC.[30] For instance, after the *Torrey Canyon* incident, the Bureau Permanent of the Comité Maritime International (C.M.I.) resolved to set up an international subcommittee to study liability problems arising from the said incident and to work in co-operation with the IMCO at the time. Replying to the questionnaire listing the issues concerning the *Torrey Canyon* incident raised in the "Documents C.M.I. 1968-No.1",[31] the Norwegian Maritime Law Association pointed out that: "...In all probability incidents of the 'Torrey Canyon' type represent a minor part of the pollution problem. The daily escape of oil and oil mixtures from thousands of ships and various kinds of installations is far more

---

28 The Merchant Shipping Act 1995 provides that the owner of a ship has strict liability for damage caused by spills of persistent oil. However, it does not impose a mandatory obligation for insurance on most ships. Only tankers and fish-factory ships are legally obliged to have insurance to enter or operate in UK waters, see Section 154. The Merchant Shipping Act 1995 is reprinted in: Hill, Christopher, *Maritime Law* (2003), Appendix 1, p. 473.

29 The OPA 90, reprinted in: DeLaRue Colin/Anderson, Charles B. *supra*, note 16, Appendix 5, p. 1005. §§2701 (37): "'vessel' means every description of watercraft or other artificial contrivance used, or capable of being used, as a means of transportation on water, other than a public vessel."

30 Wu, Chao, 'Liability and Compensation for Bunker Pollution', 33 *J. Mar. L. & Com.* 553 (2002), at 554: "Those who have followed the development of the international law of tanker pollution compensation will recall that the initial proposal for a liability regime for bunker pollution was tabled at an IMO diplomatic conference as early as 1969, during the discussion that led to the CLC..."

31 See Documentation C.M.I.1968-I. The questionnaire received replies from national associations at the time of its meeting. For the detailed replies, see Documentation C.M.I. 1968-I and Documentation C.M.I.-III.

serious."[32] This means that ships or installations other than tankers can also cause oil-pollution damage. It was furthermore considered whether the 1969 CLC should be confined to pollution by crude oil only carried as cargo or not, i.e. excluding bunkers or including bunkers. The opinions on this issue were varied. The German Maritime Law Association felt that in view of the motives behind present considerations, crude oil was the one thing that really mattered and something on which they should focus.[33] In addition, any further extension of the kinds of damage or substances to be covered would lead to great difficulties in definition as well as in consequence.[34] Therefore, after heated discussion, the 1969 CLC was framed to clearly exclude liability for bunker-oil spill in relation to the types of ships other than tankers.

The idea of a regime covering pollution from ships' bunkers was officially mooted during the discussions on the 1971 IOPC Fund and the CLC Protocol 1992. However, it was considered that to include bunkers in those instruments would complicate matters since there was a clear difference between oil carried as cargo and as bunker fuel oil.[35] However, the need for a legal regime handling liability for bunker-oil spills was considered necessary. Two main alternatives were proposed in this respect, namely: (1) a protocol to the CLC Protocol 1992 or (2) a freestanding convention.[36]

It was considered that both alternatives had significant advantages and disadvantages. On the one hand, a protocol to the CLC was simpler and faster to negotiate than a free-standing convention, as it would essentially extend the provisions of the 1992 Protocol beyond oil tankers. Nevertheless, the speed at which the protocol could be negotiated was offset by a lack of flexibility. However, a protocol would provide flexibility as to the setting of specific limits of liability without overlapping the current convention on limitation for maritime

---

32 See Documentation C.M.I. 1968-III, pp. 2-9, at 2.

33 See *ibid.*, pp. 32-41,at 34: "...It further feels that in view of the motives behind present considerations, it is only crude oil that really matters. Apart from this, any further extension of the kinds of damages or substances to be covered would lead to far-reaching difficulties in definition as well as consequences." "Crude oil" in this presentation could be understood as the oil cargo carried in the tankers.

34 For example, see *ibid.,* at 92 "The French maritime law association is in favour of the scope: the convention should cover damage by pollution of oil carried as cargo excluding bunkers." It was quite directly and obviously expressed. Also, the Yugoslav maritime law association, see *ibid.,* at 114: "we are of the opinion that all crude oil pollution damages should be covered and we would in principle prefer to include also those caused by bunker oils. The quantities carried on modern ships as bunker oil are quite substantial. However, we realise, that if bunkers are also included, the Convention would apply practically to all ships in the Worlds Merchant Marine, and we think, therefore, that these damages might be left out."

35 <http://www.imo.org/Newsroom/contents.asp?topic_id=67&doc_id=457> (visited 3 February 2005).

36 A number of delegations such as Australia, Ireland, Norway, South Africa and the United Kingdom played an active role in the IMO Legal Committee for a legal regime for dealing with bunker-oil pollution damage.

claims, i.e., the LLMC conventions.[37] A free-standing convention, on the other hand, had the advantage of dealing effectively with specific problems posed by bunkers. However, the greater complexity of a free-standing convention meant that negotiations would take longer than with a protocol, even when based on precedents established by earlier conventions. Moreover, a free-standing convention, which sets specific limits of liability in respect of bunker claims, would conflict with the LLMC Convention.[38]

Eventually, it was recognised that the existing instruments mainly dealt with oil cargo, while bunkers raised some distinctive issues that needed separate considerations and solutions. An amendment to the CLC was thus rejected and a free-standing instrument was therefore welcomed.

The matter was once again brought to great public attention in 1994, when Australia submitted a paper to the IMO Marine Environment Protection Committee and proposed a bunker pollution convention. This proposal was referred to the IMO Legal Committee, which had touched upon the issue during the negotiations on the development of the International Convention on Liability and Compensation for Damage in Connection with the Carriage of Hazardous and Noxious Substances by Sea (HNS Convention). A proposal to include bunker fuel oils was rejected on the understanding that the loophole left by this omission would be filled as soon as possible by a separate treaty instrument. According to the IMO, "Bunkers were left out of this convention, but with the firm understanding of several delegations that a bunkers' convention would be developed at the earliest possible opportunity thereafter."[39] Australia raised the issue twice again afterwards during the sessions of the IMO Legal Committee[40] before its proposal for a bunker oil convention was finally taken into consideration.

---

[37] The 1976 Convention on Limitation of Liability for Maritime Claims, as well as its 1996 Protocol, reprinted in: *Lloyd's Shipping Law Library: The Ratification of Maritime Conventions* (2004), Vol. 2, II 2.330 and II.2.340. They will thereafter be called LLMC conventions or LLMCs in this research. The LLMC conventions will be discussed in detail in Chapter 5.

[38] See IMO LEG 77/6/2.

[39] The information is available at: <http://www.imo.org/Newsroom/contents.asp?topic_id=67&doc_id=457> (visited 19 May 2005).

[40] Australia tried in October 1995 at the 73rd Session of the IMO Legal Committee and in October 1996 at the next 74th Session of the Legal Committee. The latter was a joint submission by Australia, Canada, Finland, Norway, South Africa, Sweden and the United Kingdom.

III. Risk and technical considerations[41]

From a technical point of view, the problem of bunker oil pollution is especially worthy of attention. There was no specified technical and scientific work on bunker oil spills along with the advent of the Bunkers Convention itself. However, lots of studies have been carried out in relation to the characteristics and particular challenges bunker-oil spills can cause to the marine environment.

The features of heavy fuel oil are included in the reasons why spills of bunker fuel from non-tankers increasingly become the focus of attention around the world. Heavy fuel oil is characterised by high specific gravity. Evaporation, dispersion and other natural removal processes that are often useful when cleaning up crude oil spills are slower when they relate to heavy fuel oil spills. Experience shows that many types of fuel oil in ships' bunkers are the most difficult oil to contain[42] because they are highly viscous and persistent. They can persist in the marine environment for long periods of time. Such types of oil have, therefore, the potential to cause widespread contamination to the sensitive marine environment.

Ships other than tankers are not blameless regarding oil pollution caused to the marine environment. For instance, although it was admitted that most of the oil coming ashore was from tankers, the final assessment of the killing of birds on the northeast coasts of Britain in early 1970 showed that the greatest known loss of seabirds from oil pollution in British waters was caused by ship's fuel oil.[43] Therefore, a reappraisal of the situation was called for. In addition, it has been estimated that on average the amount of fuel in bunkers carried by non-tankers is around 14 million tons at any given time – compared with approximately 130 million tons of oil carried as cargo on the world's seas.[44] Some bulk carriers and container ships carry more oil as bunker fuel than tankers carry as cargo.[45]

---

41 For further details of the technical aspects of bunker-oil spillages see data provided by the International Tanker Owners Pollution Federation Ltd and incorporated in a submission by the CMI to the IMO Legal Committee in 1996: Technical aspects of bunker oil spillages, particularly from non-tankers, see IMO LEG 74/4/2; and Clark, R.B., *Marine Pollution* (2001), pp. 64-97.

42 Ansell, D.V/Dicks, B., *et al.*, 'A Review of the Problems Posed by Spills of Heavy Fuel Oil', available at: <http://www.itopf.com/iosc2001.pdf> (visited 31 March 2005): "Over the past 25 years almost 40% of the 400 plus ship-source oil spills attended on-site around the world by ITOPF's technical staff have involved medium or heavy grades of fuel oil, either carried by tankers as cargo or *used by all types of ship as bunker fuel*. The high percentage is indicative of the fact that spills of fuel oils often cause cleanup problems, and give rise to claims to compensation that are out of proportion to the amount of oil spilled." See also Ian C. White, 'Factors Affecting the Cost of Oil Spills', available at: <http://www.itopf.com/costs02.PDF> (visited 31 March 2005).

43 See 'Oil Pollution a Dead Issue?' 2 *Marine Pollution Bulletin*, Issue 9, September 1971, pp. 129-130, at 130.

44 IMO Legal Committee: LEG 75/5/1.

45 See: <http://www.imo.org/Newsroom/contents.asp?topic_id=67&doc_id=457> (visited 23 September 2003). See also DelaRue Colin M. /Anderson, Charles B., *supra*, note 16, p. 263: "although only oil tankers can cause very large spills, oil tankers

Despite the unavailability of accurate data, it is possible to assume that, should a very large container ship lose her whole bunker load in an appalling accident, the resulting pollution would, in qualitative terms, equate to the loss of a fair-sized tanker which would thus involve payment on a large scale.[46] For example, among the 12 largest oil spills in Australia,[47] the government's response costs for oil tanker spills averaged US$115,000, while response costs for spills from non-tankers (generally heavy fuel oils) averaged over five times greater at US$625,000. Three of the four most expensive spills in terms of response costs were those from heavy fuel oil from non-tankers.[48] Commenting on the case of *Pallas*[49] in Germany, Stephen Lutter, head of the Northeast Atlantic Programme of the World Wildlife Fund for Nature (WWF) said that "the tremendous damage that can be caused by a limited amount of fuel oil from a cargo vessel gives an idea of the worse-case scenario to be expected in case an oil tanker runs aground in the same area."[50] The UK has also experienced incidents involving bunker-fuel spills where cost recovery for damage and clean-up operations has proven difficult, and in some cases, impossible – e.g. the Borodingskoye Polye (Shetland, 1993) and the Cita (Scilly Isles, 1997).[51]

Therefore, even compared with a high-profile tanker incident like *the Prestige*, the pollution damage which may be caused by numerous small bunker-oil spill incidents cannot be underestimated. The need for an international convention dealing with liability and compensation for bunker-oil spills was also highlighted by the UK P&I Clubs' "Analysis of Major Claims in 1993", which stated that "half of the total number of pollution claims arose from incidents involving ships not carrying oil cargo."[52] The International Tanker Owners Pollution Federation Limited (ITOPF)[53] has so far recorded 450 spills in 70 countries since 1971 to which it has attended. This includes 150 bunker spills from non-tankers.

---

are not the only ships carrying pollutants. Many bulk carriers and container ships carry bunker fuel of 10,000 tons or more, and there are large quantities than many of the world's tankers carry as cargo."

46 The same opinion was mentioned in 'Leading Article: Good Oil', *Lloyd's list,* October 14 1998.

47 See IMO LEG75/5/1. These 12 largest oil spills were according to the data probably from 1975 to 1997.

48 See IMO LEG 75/5/1.

49 <http://www.ramsar.org/w.n.waddensee_spill_bkgd.htm> (visited 19 May 2005). The fuel was spilled from this 7,997 GT Bahamian wood carrier Pallas, which drifted aground off Germany's Amrum Island in the North Sea in 1998. This incident oiled nearly 30,000 sea birds and environmental groups predicted that the spill could eventually affect more than 100,000 birds.

50 <http://www.ramsar.org/w.n.waddensee_spill_press.htm> (visited 19 May 2005).

51 See 'Consultation on Implementation of the Bunkers Convention', available at: <http://www.dft.gov.uk/stellent/groups/dft_control/documents/contentservertemplate/dft_index.hcst?n=14405&l=2> (visited 24 December 2005), at 5.

52 See UKP&I Club: Analysis of Major Claims 1993.

53 ITOPF is a non-profit organisation devoting considerable effort to a wide range of technical services, the most important of which is responding to oil spills.

There was no need to wait for a major disaster in order to establish a set of rules reacting to the above concerns.[54] Pollution from bunker oil is the major remaining gap in the whole package of IMO conventions dealing with marine pollution. Thus, there is a compelling demand to address this matter.

IV. The birth of the Bunkers Convention

As explained, the established liability conventions on oil pollution limit their application to oil tankers. The study in this respect shows the possible serious loss or damage which can be caused by a bunker-oil spill. As the result, the Bunkers Convention was adopted. The Convention, which has not yet entered into force, is a freestanding instrument and is largely modeled on the CLCs.

To a large extent, it remains to be seen whether the Bunkers Convention will benefit pollution victims of bunker-oil spills as provided in the preamble to the Bunkers Convention which is "to ensure the payment of adequate, prompt and effective compensation for damage caused by pollution resulting from the escape or discharge of bunker oil from ships". In order to see the overall picture, an overview of the Bunkers Convention is given in the section immediately following.

C. Overview of the Bunkers Convention

I. Categories of ships

1. "Ship"

The Bunkers Convention defines a "ship" as "any seagoing vessel and seaborne craft, of any type whatsoever".[55] This definition corresponds to the main objective of the Convention. Therefore, it will cover bunker-oil spills from all types of vessel. Nevertheless, such a broad definition may well cover a large number of floating objects in the sea. Accordingly, once the Bunkers Convention is ratified, it will impose various burdens on non-tanker vessels and seaborne craft.

2. Does the Bunkers Convention apply to oil tankers?

It is not the purpose of the Bunkers Convention to replace the earlier civil liability conventions. It is quite clear from Article 4(1) of the Bunkers Convention that:

---

54 During the preparatory works for the Bunkers Convention, many delegations expressed the above concern, such as the submission by the International Association of Ports and Harbours, see IMO LEG 78/5/1.

55 The Bunkers Convention, Art.1(1).

> "This Convention shall not apply to pollution damage as defined in the Civil Liability Convention, whether or not compensation is payable in respect of it under that Convention."

The "Civil Liability Convention" in the said Article alludes to the established CLCs and the Fund Conventions, which were adopted to ensure that prompt and adequate compensation is available to victims who suffer oil-pollution damage from tankers or other seagoing vessels constructed or adapted to carry cargos of oil, as described earlier in this chapter.[56] The Bunkers Convention focuses only on the liability issues arising from bunker-oil spill incidents. It does not apply to oil tankers or oil-pollution damage resulting from the escape or discharge of oil from tankers.

3. *"Warships"*

Article 4, which provides for exclusions, specifically deals with "warships" and some other ships owned or operated by States as seen in paragraphs 2, 3 and 4.

Article 4(2) of the Convention excludes "warships, naval auxiliary and other ships owned or operated by a State and used, for the time being, only on Government non-commercial service". In order to understand the preferential treatment of these types of ships listed in this article, one may go back to the preparatory work and the discussions during the conference of the 1969 CLC. With the aim of obtaining a sufficient number of ratifications for the civil liability convention, a compromise that the convention would not apply to those types of ships was agreed upon in the 1969 CLC.[57] It was possibly for the same reason that a similar provision in the Bunkers Convention was reproduced therein.

However, if the State-owned vessel is used for commercial purposes, Article 4(4) shall apply. It provides that the State-owned vessel would be treated in the same way as any other types of vessel. To quote:

> "With respect to ships owned by a State Party and used for commercial purposes, each State shall be subject to suit in the jurisdictions set forth in Article 9 and shall waive all defences based on its status as a sovereign State."[58]

Article 4(3), on the other hand, contains a provision which states that:

> "A State Party may decide to apply this Convention to its warships or other ships described in paragraph 2, in which case it shall notify the Secretary-General thereof specifying the terms and conditions of such application."[59]

---

56 See section A.II. of this chapter.
57 Wu, Chao, *supra*, note 3, p. 41.
58 The Bunkers Convention, Art. 4(4).
59 The Bunkers Convention, Art. 4(3).

II. "Oil"

The Bunkers Convention will not be applicable unless the vessel concerned is carrying bunker oil and the latter caused pollution damage. Bunker oil is defined as "any hydrocarbon mineral oil, including lubricating oil, used or intended to be used for the operation or propulsion of the ship, and any residues of such oil."[60]

The definition of "oil" denotes the presence of the following three elements:

(i) any hydrocarbon mineral oil, including lubricating oil;
(ii) the oil falling within (i) is used or intended to be used for the operation or propulsion of the ship;
(iii) any residues of such oil.

One may notice that it does not contain the word "persistent" in the definition, while "persistent" is present in the "oil" definition of the 1969 CLC and its 1992 Protocol. "Oil" in the CLC Protocol 1992 means: "any persistent hydrocarbon mineral oil such as crude oil, fuel oil, heavy diesel oil and lubricating oil, whether carried on board a ship as cargo or in the bunkers of such a ship."[61] The CLC conventions stress the "persistent" nature of the oil carried as cargo in tankers. They deal solely with persistent oils since there is a possibility that a tanker carries a type of non-persistent oil and light products as cargo.[62] Non-persistent oil evaporates quickly and causes much less and nearly no damage to the marine environment. Compared with the variety of oil cargoes carried on tankers, bunker oil, on the other hand, may be expected to persist due to its greater proportion of non-volatile components and its high viscosity.[63] However, "any hydrocarbon mineral oil" in the Bunkers Convention is broad enough to include both "persistent" and "non-persistent" hydrocarbon mineral oil, although the latter is rarely in the nature of any type of bunker fuel oil.[64]

The Bunkers Convention is the convention relating to oil carried as fuel oil used for the operation or propulsion of a ship. The determination of whether "oil" is "used or intended to be used for the operation or propulsion of the ship" depends on the facts in a particular incident, where technical experts may be

---

[60] The Bunkers Convention, Art.1(5).

[61] The CLC Protocol 1992, Art.2(2); see also the 1969 CLC, Art. 1(5): "'Oil' means any persistent oil such as crude oil, fuel oil, heavy diesel oil and lubricating oil, whether carried on board a ship as cargo or in the bunkers of such a ship."

[62] For further discussion regarding "persistent" oil, see Abecassis, D. W. (ed.), *supra*, note 5, pp. 196-197.

[63] See the paper completed by the ITOPF with the submission of CMI in the 74th session of IMO legal committee, see IMO LEG 74/4/2 for more information.

[64] It might have been done by the drafters of the Bunkers Convention intentionally. See IMO Publication: *Manual on Oil Pollution: Section IV-Combating Oil Spills*, at 8: "Whilst the term persistent is not precisely defined in any convention or international standard, generally, oils that are normally termed persistent include crude oils, fuel oils, heavy diesel and lubricating oils. Non-persistent oils include gasoline, light diesel oil and kerosene."

needed. If the ship is carrying oil as cargo[65] at the time of the incident and not for the "operation or propulsion of the ship", the Bunkers Convention will be inoperative. As noted earlier in this chapter,[66] if the ship in question is actually "constructed or adapted for the carriage of oil in bulk as cargo", the CLCs and the Fund Conventions will govern liability and compensation whether the oil is carried as cargo or fuel.

Bunker-oil pollution damage may occur during the bunkering process when fuel oil is transferred from one supplying ship to another cargo ship or even, for instance, to a fishing boat at times. The International Chamber of Shipping noted that only the damage originating from fuel on board a ship should be covered by the proposed bunkers convention.[67] In accordance with the current provisions in the Bunkers Convention, it is understood that "bunker oil" includes both the oil that is used and the oil that will be used for the operation or propulsion of the ship. The time factor is thus irrelevant. However, we must specify some circumstances: if the supplying ship is an oil tanker or oil barge, it may be necessary to examine whether the escape or discharge of oil occurred from the ship taking the bunker fuel or whether it occurred from the vessel supplying it. If the oil escaped from the oil tanker supplying the oil, the Bunkers Convention is not applicable; the consequent issues of liability or compensation will therefore fall within the scope of the CLCs. However, if the spill or escape of fuel oil was from a cargo ship, which was intended to transport after the bunkering operation, the Bunkers Convention will apply since the oil was to be used for the operation or propulsion of the ship.

The Bunkers Convention also covers pollution damage caused by any residues of "bunker oil", i.e. "any hydrocarbon mineral oil, including lubricating oil, used or intended to be used for the operation or propulsion of the ship". It usually happens after a sea transport.

III. Scope of application

Article 2 of the Bunkers Convention, which defines the scope of application, obviously follows the format of the CLC Protocol 1992. It defines the scope of the Bunkers Convention and states that:

> "This Convention shall apply exclusively:
> (a) to pollution damage caused:
> (i) in the territory, including the territorial sea, of a State Party, and
> (ii) in the exclusive economic zone of a State Party, established in accordance with international law, or, if a State Party has not established such a zone, in an area beyond and adjacent to the territorial sea of that State determined by that State in accordance

[65] Usually, these oil cargoes will be for other purposes. Even if they are fuel oil and can be for the "operation or propulsion of the ship", they cannot be regarded as "bunker oil" in the context of the Bunkers Convention.

[66] See Section B.I of this chapter.

[67] See IMO LEG 74/4/4.

with international law and extending not more than 200 nautical miles from the baselines from which the breadth of its territorial sea is measured.
(b) to preventive measures, wherever taken, to prevent or minimise such damage."[68]

Accordingly, the application of the Convention therefore depends on the place of damage, which must fall within the geographical sphere defined by the Convention. The place where the oil actually escaped is of little importance. The Convention may apply to the incident in which oil is spilt on the high seas, provided that pollution damage is sustained within the territorial sea or the exclusive economic zone of a State Party.

The second part of Article 2 provides that the measures taken to prevent or minimise pollution damage are not limited to the geographical area defined in the first part. The argument in support of this point as regards tanker-oil pollution incidents is also helpful for understanding similar issues in bunker-oil spill incidents. It has been pointed out that: "if the incident took place on the high seas, it is usual to take preventive measures at the place of incident in order to prevent the pollution from spreading into territorial waters."[69] The costs involved in undertaking preventive measures, wherever made, will accordingly fall within the ambit of the Convention.

The Bunkers Convention deals with liability for pollution damage caused by oil when carried as fuel in non-tanker vessels. The "pollution damage" is defined in Article 1(9) to mean:

"(a) loss or damage caused outside the ship by contamination resulting from the escape or discharge of bunker oil from the ship, wherever such escape or discharge may occur, provided that compensation for impairment of the environment other than loss of profit from such impairment shall be limited to costs of reasonable measures of reinstatement actually undertaken or to be undertaken, and
(b) the cost of preventive measures and further loss or damage caused by preventive measures."

The definition of "preventive measure" in Article 1(7) of the Bunkers Convention is: "…any reasonable measure taken by any person after an incident has occurred to prevent or minimise pollution damage." The definition is applicable even if the oil was not actually spilt, since an "incident" in the context means "…any occurrence or series of occurrences having the same origin, which causes pollution damage or creates a grave and imminent threat of causing such damage."[70] The "incident" must have actually occurred according to the 1969 CLC.[71] However, the CLC Protocol 1992 recognised the drawbacks in this definition and included the incident that "...creates a grave and imminent threat of causing such

[68] The Bunkers Convention, Art. 2.
[69] Wu, Chao, *supra*, note 3, p. 45.
[70] The Bunkers Convention, Art. 1(8).
[71] The 1969 CLC, Art. I(8): "'Incident' means any occurrence, or series of occurrences having the same origin, which causes pollution damage."

damage".[72] Obviously, the Bunkers Convention follows the definition of "preventive measures" in the CLC Protocol 1992. In effect, the "incident" definition which includes "grave and imminent threat of causing such damage" was firstly discussed in and adopted by the 1984 Protocol to amend the International Convention on Civil Liability for Oil Pollution Damage, 1969.[73] Professor Z. Brodecki[74] pointed out that:

> "...the discussion concentrated on the qualification of the word "threat" in the definition of 'incident', by 'serious' or 'grave and imminent'. 'Grave and imminent' appeared to be more appropriate, because it was used by the International Convention Relating to Intervention on the High Seas (Art. I) and UNCLOS (Art.221). Although there was strong support for 'serious', the words 'grave and imminent' were accepted by the Protocol to the International Convention on Civil Liability for Oil Pollution Damage 1969."[75]

The preventive measures must be taken to prevent pollution damage; otherwise the costs of the operation would not be covered by this definition. It is important to distinguish "preventive measures" from "salvage". The purpose of the latter is to salve hull or cargo and the hull insurer of the vessel pays for the costs of the salvage operation. Salvage operations are regulated by the 1910 Salvage Convention, which was replaced by the 1989 London Convention.[76] In the event of a tanker oil spill, if the operations had a dual purpose, i.e. both to prevent and minimise pollution and save the vessel and cargo, it was necessary to distribute the costs of operations of salvage and pollution prevention. Such a distribution depends upon the assessment of facts in a specific case.[77]

---

72 The 1992 CLC, Art.2(4): "'Incident' means any occurrence, or series of occurrences having the same origin, which causes pollution damage or creates a grave and imminent threat of causing such damage."

73 The 1984 Protocol, Art.2(4), see the above Section A.II of this chapter: the 1984 Protocol did not enter into force.

74 Professor Z. Brodecki was at the time the professor of Law and Head of the Maritime Law Department, the University of Gdansk. He had authored Polish proposals regarding the definition of pollution damage, and attended as a representative of Poland the International Conference on Liability and Compensation for Damage in Connection with the Carriage of Substances by Sea, held in London in 1984. That conference adopted (i) the Protocol of 1984 to amend the International Convention on Civil Liability for Oil Pollution Damage 1969, and (ii) the Protocol of 1984 to amend the International Convention on the Establishment of an International Fund for Compensation for Oil Pollution Damage 1971.

75 Brodecki, Z., 'New Definition of Pollution Damage', *Lloyd's Mar. & Com. L.Q.* (1985), pp. 382-391, at 390.

76 The International Convention on Salvage, 1989 was adopted in London in April 1989 under the auspices of the International Maritime Organization and it has entered into force in July 1996. The full text of the convention is available at: <http://www.admiraltylawguide.com/conven/salvage1989.html> (visited 25 December 2005).

77 See Wu, Chao, *supra*, note 3, pp. 281-288, for more discussion about the distinction between "salvage" and "preventive measures".

A uniform interpretation of the definition of "pollution damage" is essential for the operation of a regime of compensation established by the international civil liability convention. Further analysis of pollution damage will be given in Chapter 6 of this research.[78]

IV. Liability established by the Bunkers Convention

1. *Liable parties*

In order for the pollution victim to identify the liable person, the Bunkers Convention attributes liability to the shipowner,[79] a term which includes the "registered owner, bareboat charterer, manager and the operator of the ship".[80] With the said provisions, two crucial points were acknowledged, *viz*: first, increasing the number of liable persons in the definition of "shipowner" might expand the recovery of compensation for pollution victims; secondly, care must be taken to ensure that no overlapping insurance would be taken out by the liable persons.[81]

A number of proposals for the definition of "shipowner" were tabled by the delegations to the IMO Legal Committee meetings.[82] These proposals were considered during the drafting of a bunkers convention. For example, in the 80th session of the Legal Committee, several delegations came up with some draft articles for the proposed convention containing two options: "(1) the shipowner means the owner, including the registered owner, bareboat and demise charterer, manager and operator of the ship; (2) the shipowner means the person or persons registered as the owner of the ship, or in the absence of registration, the person or persons owning the ship. However, in the case of a ship owned by a State and operated by a company which is registered as the ship's operator in that State, the "owner" shall mean such company."[83]

The first option was based on Article 1(2) of the 1976 International Convention on Limitation of Liability for Maritime Claims (1976 LLMC). It was widely recognised that the Bunkers Convention should entitle the shipowner to limit liability in accordance with applicable national or international law, and in most cases that meant that the 1976 LLMC as well as its 1996 Protocol would apply.[84]

---

78 See Chapter 6, Section E.

79 The Bunkers Convention, Art. 3(1).

80 The Bunkers Convention, Art. 1(3).

81 See IMO LEG 80/4/1: it is also recorded that P&I Clubs have difficulties in providing cover for a vessel where the liability is jointly and severally held by all the persons defined as owner, especially if the Club does not know all those persons.

82 See IMO LEG 78/WP.3, agenda item 5 citing document IMO LEG 78/5/2.

83 See IMO LEG 80/4/1.

84 The Bunkers Convention, in Art.6, entitles the shipowners to limit liability under any applicable national or international regime, such as the 1976 LLMC, as amended.

Therefore, in order to ensure uniformity, a similar definition of "shipowner" as the one in the LLMCs, i.e. "owner, charterer, manager and operator of a seagoing ship", was considered in the Bunkers Convention. The same definition was deemed advantageous for the application of relevant limitation conventions. The second option was identical to the provision in the 1992 CLC due to the Bunkers Convention being largely modelled on the CLC Conventions.[85]

After a long and intensive discussion, the first option, with a few minor revisions, was adopted. The "shipowner" including "registered owner, bareboat charterer, manager and the operator of the ship" shall be liable for pollution damage at the time of the incident. The "registered owner" is separately defined as:

> "...the person or persons registered as the owner of the ship or, in the absence of registration, the person or persons owning the ship. However, in the case of a ship owned by a State and operated by a company which in that State is registered as the ship's operator, 'registered owner' shall mean such company."[86]

This definition is similar to the definition of "owner" in the CLC Conventions.[87] Furthermore, with respect to the incident consisting of a series of occurrences having the same origin, the liability shall attach to the shipowner "at the time of the first of such occurrence."[88] In addition, the liability is imposed on the owner of a ship "at the time of an incident" from which pollution results.[89] This would, however, result in the impossibility of finding the liable person in cases where the oil spill source is untraceable.

2. *Channelling of liability*

Channelling of liability is the mechanism, which directs all claims to a specific person. It is utilised in the CLCs. Precisely, under the CLC Protocol 1992, the liability for pollution damage is narrowly channelled through a long list to the owner of the ship, the owner under the CLCs meaning the registered owner only.[90]

---

85 See IMO LEG 80/4/1.

86 The Bunkers Convention, Art.1(4).

87 The 1969 CLC, Art. I(3).

88 The Bunkers Convention, Art. 3(1).

89 The Bunkers Convention, Art. 3(1).

90 The CLC Protocol 1992, Art.4(2): "No claim for compensation for pollution damage may be made against the owner otherwise than in accordance with this Convention... no claim for compensation ... may be made against: (a) the servants or the agents of the owner or the members of the crew; (b) the pilot... (c) any charterer...manager or operator of the ship;(d) any person performing salvage operations... (e) any person taking preventive measures; (f) all servants or agents of persons mentioned in subparagraphs (c),(d) and (e); unless the damage resulted from their personal act of omission, committed with the intent to cause such damage, or recklessly and with knowledge that such damage would probably result." Comparatively, the channelling mechanism in the 1969 CLC did not succeed in directing all actions for pollution damage towards the shipowner, see the 1969 CLC, Art. III(4), "No claim for compensation for pollution damage shall

This channelling mechanism thus directs all actions for pollution damage towards the registered owner. Meanwhile, the owner retains the right to exercise his recourse action against any third parties at fault.[91] It was pointed out that "...channelling of liability can be a useful and desirable tactic."[92] The merits are not described completely here. It at least puts a named person on a clear notice of potential liability, thereby inducing him to contract for insurance covering the risk involved.[93] In addition, the chemical and physical nature of oil necessitates a rapid response; the channelling mechanism is particularly helpful in speeding up response actions and therefore eliminates pollution damage in scope and extent.

During the discussions of a bunkers convention, there were different viewpoints on the subject of channelling of liability. One main suggestion was that the liable persons should be as close as possible to the operation of the ship, and thus the liability for damage caused by the ships' bunker fuel should attach not only to the registered owner but also to a small group of parties responsible for the day-to-day operation of the ship, i.e. the charterer, manager or operator.[94] Admittedly, the owner, operator, manager and bareboat charterer under the bareboat charter-party may all be involved in operations of the ship or have actual custody of the ship. Placing liability on those parties may thus provide an incentive for them to take good precautionary measures and to minimise pollution damage by responding to an oil spill as soon as possible. It would be fair to hold liable the charterer, manager or operator of a ship that caused a bunker spill, rather than the shipowner alone, in cases where the latter had taken all reasonable measures to maintain the ship in question.[95] However, if the liability were imposed on all these relevant parties, it would increase the difficulty of apportionment of liability and eventually delay the response and compensation.

The International Group of P&I Clubs argued from the insurance point of view and proposed another alternative. In practice, different parties and their insurers are unlikely to agree quickly on apportionment of liability. The Club therefore suggested that: "this security (recovery) may be provided not by making all parties liable, as in the present draft, but instead by making the other parties liable only

---

be made against the owner otherwise than in accordance with this Convention. No claim for pollution damage under this Convention or otherwise may be made against the servants or agents of the owner."

91 Wu, Chao, *supra*, note 3, p. 171. The parties (for example, the servants or agents of the owner or the members of the crew, the pilot, any charterer, manager or operator of the ship and so on who are on the list in Article 4(2) of the CLC Protocol 1992) are protected by channelling in the absence of wilful misconduct. They will not be primarily liable towards claimants, but can be secondarily liable to the shipowner in an action for recourse.

92 See IMO LEG 77/4/3.

93 See *ibid.*

94 See IMO LEG 74/4/1 citing the IMO Legal Committee's 73rd session. United States submitted a separate document on this issue: IMO LEG 77/4/3.

95 See IMO LEG 77/4/3. In a submission by the United States, it describes the experience of the United States in successfully channelling pollution liability to a small group, rather than channelling solely to a single party.

when it has not been possible to obtain compensation from the registered owner or his insurer."[96] The International Chamber of Shipping, sharing the same view, added that "...For practical reasons and reasons of consistency with other pollution conventions, proper channelling provisions should be included in the proposed convention."[97]

In the absence of a fund similar to the IOPC Fund in providing a second-tier compensation, the reason for not having adopted a similar channelling mechanism in the Bunkers Convention is to reassure claimants that adequate compensation will be possible, even if the registered owner is not able to pay for the pollution damage or has paid out of his own pocket.[98] At the same time, there is a limitation on the registered owner's liability. It is difficult to evaluate whether the compensation limits provided by the Bunkers Convention will be sufficiently high. To some extent, this non-channelling mechanism may be a more reliable means for sufficient compensation, although it will create difficulties in settling claims.

The Bunkers Convention eventually adopted the approach that encourages pollution victims to pursue a range of persons defined as "shipowner" in Article 1(3) of the Bunkers Convention.[99] It is an obvious shift away from the mechanism of channelling. The effect of such an approach still remains to be seen. Since liability and its channelling relate to the insurance aspect, more related discussions are, therefore, necessary.[100]

3. Shipowners' liability is joint and several

Article 3 of the Bunkers Convention contains a provision to the effect that where more than one person is liable, the liability shall be joint and several among them, since it states: "when more than one person is liable in accordance with paragraph 1, their liability shall be joint and several." The concept of "person" is defined by the Convention as "any individual or partnership or any public or private body, whether corporate or not, including a State or any of its constituent subdivision."[101] According to Article 3 (1), "person" is understood as the shipowner as defined, since the liability is imposed on the shipowner.

The "joint and several liability" exists not only among the liable persons in one incident, but also among ships when an incident involves two or more ships. Article 5 provides for a joint and several liability when two or more ships were involved in an oil-spill incident. It states:

> "When an incident involving two or more ships occurs and pollution damage results therefrom, the shipowners of all the ships concerned, unless exonerated under Article 3,

[96] See IMO LEG/CONF.12/9.
[97] See IMO LEG/CONF.12/10.
[98] Wu, Chao, *supra*, note 30, at 559.
[99] Shipowner includes "registered owner, bareboat charterer, manager and operator of the ship".
[100] See Chapter 3 of this research.
[101] The Bunkers Convention, Art. 1(2).

shall be jointly and severally liable for all such damage which is not reasonably separable."

This Article applies only if all vessels concerned are ships within the meaning of the Bunkers Convention, which includes "any seagoing vessel and seaborne craft, of any type whatsoever".[102]

Apparently, the purpose of these articles is to hold all persons and ships liable when two or more ships are involved. It is reasonable, since if one of them is unfortunately insolvent, the others must therefore pay the whole costs of pollution damage claimed under the Convention.

4. *The basis of liability and exonerating circumstances*

Liability under the CLCs is by nature a strict liability. This was one of the most debated issues from the outset of the preparatory work to the completion of the 1969 CLC. The strict liability or no-fault liability means that the shipowner is liable irrespective of the existence of any fault on his side. In other words, he is liable simply because of the fact that his ship carrying persistent oil had a spill and caused pollution damage. Under the CLCs, the claimants need only prove that the damage sustained by them was caused by an oil-spill incident in question. There were several reasons advanced against strict liability at the time of the discussion of an international instrument.[103] Two main reasons were: first, the concept of strict liability was foreign to the law of the sea at the time and its introduction would upset the whole structure of private maritime law. The problem of oil pollution was something which did not call for an exceptional treatment. Second, the victims of pollution from oil which has escaped from a ship should not be

[102] The Bunkers Convention, Art. 1(1).

[103] For example, in Documentation C.M.I.1968-I, at 112, the Report of the International Subcommittee: "The arguments advanced against imposing liability without fault include the following: 1. It would be inequitable to give those sustaining pollution damage a preferred status vis-à-vis personal injury, death and property damage claimants with claims arising out of other marine casualties. 2. A seaworthy steamship or motor vessel, properly manned, is not per se a dangerous instrumentality, and the operator should not be required to pay for or insure against losses not caused by his fault...3. Oil cargoes are not per se dangerous, and their owners should not be required to pay for or insure against losses caused by such cargoes. 4. In a competitive market, it may not be possible for the ship-owner to pass on the extra cost of insurance against liability without fault to the shipper in the form of additional freight; shipowners large enough to be self-insurers may have an advantage over small operators. Furthermore, the owner of tankers under long term time or consecutive voyage charter would be unable to increase the charter rate or hire or freight so as to recover the cost of insuring against liability without fault. 5. the Liberian Board of Investigation found the '*Torrey Canyon*' at fault. Hence, no need for a convention imposing liability without fault is indicated by the case which prompted this study."

given preferential treatment as compared with the victims of other risks.[104] Eventually, a final compromise was reached. According to one author:

> "It was not until the final days of the Conference that the negotiators were able to find a compromise due to concessions on the part of the U.K. delegation in withdrawing their opposition to strict liability, accepting a maximum insurance limit on the insurance market, and another convention to examine the constitution of a fund to complement the shipowner's strict liability."[105]

Apparently, the establishment of strict liability was sensitively related to the issue of insurance and the limitation of liability. The same is true of the provisions of the Bunkers Convention. The interactions among them will be detailed in the following chapters of this research.

The Bunkers Convention virtually establishes the same strict liability principle as contained in the 1969 CLC and its 1992 Protocol. Article 3(1) of the Bunkers Convention provides that:

> "Except as provided in paragraphs 3 and 4, the shipowner at the time of an incident shall be liable for pollution damage caused by any bunker oil on board or originating from the ship, provided that, if an incident consists of a series of occurrences having the same origin, the liability shall attach to the shipowner at the time of the first of such occurrence."

Accordingly, the shipowner is strictly liable for pollution damage caused by any bunker oil spilt from his ship, unless it is established that: a) the damage resulted from an act of war or natural phenomenon; b) the damage was wholly caused by an act or omission done with the intent to cause damage by a third party; c) the damage was wholly caused by the negligence or other wrongful act of any Government or other authority in relation to maintaining navigational aids.[106]

The shipowner may also be wholly or partially exempt from liability under Article 3(4), which provides:

> "If the shipowner proves that the pollution damage resulted wholly or partially from an act or omission done with intent to cause damage by the person who suffered the damage or from the negligence of that person, the shipowner may be exonerated wholly or partially from liability to such person."

The preceding Article is identical to Article III(3) of the 1969 CLC, and the same was followed in the CLC Protocol 1992.[107] If the fault of the victim is the sole cause of or contributory to the pollution damage, the shipowner will be wholly or partially discharged from his liability, according to the extent of fault of the victim in causing the damage.

---

[104] See Documentation C.M.I.1968-III, pp. 145-146.

[105] Wu, Chao, *supra*, note 3, pp. 58-59, the reference is from: O.R. 1969, LEG/CONF/C.2./WP.35, November 24, 1969, pp. 596-597.

[106] The Bunkers Convention, Art. 3(3).

[107] Abcassis, David W. (ed.), *supra*, note 5, p. 206. Under the CLCs, the provision was introduced on the ground of fairness, despite anxieties on the part of some that it would weaken the concept of strict liability. Any footnotes omitted.

Article 3 also contains a provision to the effect that claims for bunker-fuel pollution damage can only be brought against the shipowner under the Convention when it is applicable, and claimants cannot turn to other national or international legislations.[108]

Shipowners, however, may have the right of recourse independently of the Bunkers Convention. The right of the shipowner to recover from third parties is expressly preserved by the Convention.[109]

5. *Limitation of Liability*

Under the principles of maritime law, the right to limit liability is quite typical. Similarly, the drafters of the 1969 CLC and the 1971 Fund Convention unanimously agreed to make it possible for the liability for pollution damage to be limited. The reasons were summarised as follows: (1) As a logical and a practical matter, it would be impossible to abolish the limitation of pollution liability and yet maintain the principle in respect of other liabilities, notably the liability for death and personal injury; (2) the main reason for retaining the principle is that the insurance for the shipowner's liability risk is greatly facilitated when the maximum liability can normally be assessed in advance, and it is also in the interest of the victims that the insurance for adequate compensation is easily available.[110]

The Bunkers Convention maintains this tradition of maritime law. In the preamble to the Bunkers Convention, it emphasises "the importance of establishing strict liability for all forms of oil pollution which is linked to an appropriate limitation of the level of that liability." An article on "Limitation of liability" was thus provided in the Bunkers Convention.[111] However, although the Bunkers Convention entitles the shipowner to limit his liability, the relevant provision regarding limitation of liability is too general compared with the similar provisions in the CLCs.[112] In comparison with the long and detailed provisions of the limitation of liability in the CLCs, Article 6 in the Bunkers Convention only stipulates that:

> "Nothing in the Convention shall affect the right of the shipowner and the person or persons providing insurance or other financial security to limit liability under any applicable national or international regime, such as the Convention on Limitation of Liability for Maritime Claims, 1976, as amended."

---

[108] The Bunkers Convention, Art. 3(5): "No claim for compensation for pollution damage shall be made against the shipowner otherwise than in accordance with this Convention."

[109] The Bunkers Convention, Art. 3(6): "Nothing in this Convention shall prejudice any right of recourse of the shipowner which exists independently of this Convention."

[110] See Documentation C.M.I. 1968-III, Report submitted by International Subcommittee, at 146.

[111] The Bunkers Convention, Art. 6.

[112] The 1969 CLC, Art. V, and the CLC Protocol 1992, Art. 6.

This confirms the shipowner's right to limit his liability. However, the extent of limitation relies on "any applicable national or international regime". Apparently, this article is not intended to establish a separate limitation regime or a limitation fund available to be exclusively devoted to bunker-fuel pollution claims. The claimants for bunker-fuel pollution damage will have to compete with other property claimants in the same incident. It also appears from this article that the shipowner may seek to limit his liability under the provisions of the 1976 LLMC or its 1996 Protocol, if they are applicable; or otherwise other applicable national legislations in this respect. The implication of this issue will be given a more detailed explanation in Chapter 5 of this research.

It should be noted that at the conference which adopted the Bunkers Convention, three other resolutions were also adopted.[113] One of these was the resolution on the limitation of liability. This resolution urges all States that have not yet ratified the Protocol of 1996 to amend the Convention on Limitation of Liability for Maritime Claims, 1976 (LLMC Protocol 1996), or accede thereto to finally ratify, or accede to the Protocol.[114] This resolution, to some extent, tries to make the applicable limitation rule for liability certain and uniform when implementing the Bunkers Convention. Moreover, in particular, compared with the 1976 LLMC, the LLMC Protocol 1996 raises the limits of liability and therefore increases the amounts of compensation payable in the event of an incident. It may thus somehow benefit pollution claimants. Furthermore, the States are also encouraged to denounce the Limitation Conventions of 1924 and 1957.[115] Denunciation was not, however, made a prerequisite for becoming a party to the Bunkers Convention for ethical considerations.

V. Compulsory insurance and direct recourse

The principle of compulsory insurance is not unique to the Bunkers Convention, as it has become one of the general features of liability conventions.[116] The formula for compulsory insurance in Article 7 (1) of the Bunkers Convention, which is quite similar to other civil liability conventions, states that:

> "The registered owner of a ship having a gross tonnage greater than 1 000 registered in a State Party shall be required to maintain insurance or other financial security, such as the guarantee of a bank or similar financial institution, to cover the liability of the registered owner for pollution damage in an amount equal to the limits of liability under the applicable national or international limitation regime, but in all cases, not exceeding an

[113] More see Section VIII of this chapter.

[114] The other two resolutions are: (1) Resolution on the promotion of technical co-operation; (2) Resolution on protection for persons taking measures to prevent or minimise the effects of oil pollution.

[115] See IMO LEG 82/3/3. More discussion regarding the Limitation Conventions of 1924 and 1857 will be given in Chapter 5, Section B.I.

[116] For example: CLCs, the HNS Convention of 1996.

amount calculated in accordance with the Convention of Limitation of Liability for Maritime Claims,1976, as amended."[117]

Accordingly, the Bunkers Convention requires that ships which have the significant potential to cause pollution damage be covered by insurance.

Compulsory insurance is only imposed on the ships registered in the State Party. There is thus a worry that low-standard ships will register with non-party States. However, according to Article 7(12) of the Bunkers Convention, each State Party shall ensure that, under its national law, insurance or other financial security is in force in respect of *any ship* having a gross tonnage greater than 1,000, *wherever registered*, entering or leaving a port in its territory, or leaving an offshore facility in its territorial sea.[118] Therefore, it may solve the problem of unfair competition between a State Party and a non-State Party. According to this paragraph, ships having a gross tonnage greater than 1,000, whether registered in a State Party or not, shall take out the same level of insurance in order to enter the port of a State Party.[119] Nevertheless, if ships registered in a non State Party are not intended to transport into the territory of a State Party, it is under their discretion to purchase insurance.

1. Three prerequisite factors

There are three relevant phrases in the aforesaid provision which are critical to proper operation or implementation of liability insurance. They are "registered owner of a ship", "gross tonnage greater than..." and "in an amount equal to the limits of liability..."

a) "Registered owner of a ship"

It should be noted that the duty to obtain insurance is imposed only on the registered owner to the exclusion of others who come within the definition of liable party – "shipowner" in Article 1(3) of the Bunkers Convention. The "registered owner" is defined in Article 1(4) of the Bunkers Convention.[120]

---

117 The Bunkers Convention, Art. 7(1).

118 The Bunkers Convention, Art.7(12): "Subject to the provisions of this Article, each State Party shall ensure, under its national law, that insurance or other security, to the extent specified in paragraph 1, is in force in respect of any ship having a gross tonnage greater than 1000, wherever registered, entering or leaving a port in its territory, or arriving at or leaving an offshore facility in its territorial sea.

119 According to the Convention, it is obvious from Article 7(1) that a ship in a State Party shall be required to purchase insurance. In addition, according to the meaning of Article 7(12), any ship, even if it does not belong to a State Party, once it wants to enter a port of State Party, has to take out insurance.

120 The Bunkers Convention, Art. 1(4): "'registered owner' means the person or persons registered as the owner of the ship or, in the absence of registration, the person or persons owning the ship. However, in the case of a ship owned by a state and operated by a company which in that State is registered as the ship's operator, 'registered owner' shall mean such company."

When the bunkers convention was negotiated, different views were presented by the delegations to the conference. The International Chamber of Shipping argued that the very nature of the CLC and HNS cargo[121] posed dangers and that the cargo interests involved should be called upon to contribute in the event of an incident. In the conventions regarding those types of cargoes, liability is, for the sake of expediency, channelled to the registered owner and the existence of cargo interests' funds ensures that the claimants are adequately compensated. However, incidents involving ships' bunkers are more likely to be linked to operation of the ship rather than to the nature of the substance itself. It might, therefore, be more appropriate for liability to attach to the responsible party rather than be channelled to the registered owner,[122] since the person who is registered as the owner is not necessarily the one operating the vessel concerned.

From the insurance point of view, imposing the obligation only on the registered owner is unfair, since the liability insurance taken out by the registered owner will not necessarily cover the liabilities of other parties. As discussed in an earlier section in this chapter, the Bunkers Convention does not follow the notable liability-channelling provisions in the CLCs and places liability on multiple parties within the "shipowner" definition.[123] The insurance market would however have no difficulty in providing separate insurance cover for those persons.

In practice, it is not always easy to discover the liable person. In order to simplify the process of identification, the Bunkers Convention finally imposes the obligation to take out insurance or other financial security on the registered owner. The ship's registration is an administrative act, by which the nationality and collateral rights and duties are conferred on a ship. The basic matters in relation to the ship are thus entered in the public records.[124] It is relatively easier for claimants to identify the registered owner of a ship and therefore find the liable person. However, it would lead pollution victims in the first instance to pursue the registered owner or his liability insurer and this cannot satisfy the registered owner per se.

b) "Gross tonnage" – insurance threshold

"Gross tonnage" refers to the gross tonnage calculated according to the International Convention on Tonnage Measurement of Ships, 1969.[125] Annex I of the

[121] HNS cargo is intended to be regulated under the International Convention on Liability and Compensation for Damage in Connection with the Carriage of Hazardous and Noxious Substances, adopted by the IMO in 1996, which has not come into force. The convention is reprinted in: *Lloyd's Shipping Law Library: The Ratification of Maritime Conventions* (2004), Vol.4, II.7.125.

[122] See IMO LEG 74/4/4.

[123] See section C.IV.2 of this chapter.

[124] See Coles, Richard/Ready, Nigel, *Ship Registration: Law and Practice* (2002), Özçayir, Z. Oya, *Port State Control* (2001), pp. 10-12, for more details about ship's registration.

[125] This convention was adopted by the International Maritime Organization in 1969 and entered into force in July 1982, reprinted in: *Lloyd's Shipping Law Library:*

said convention contains details of how to calculate a ship's gross tonnage and net tonnage. The amount of gross tonnage is closely related to the insurance threshold, i.e., to what extent of gross tonnage the ship is required to purchase insurance.

Three options by which the insurance threshold was to be decided were proposed during the negotiations. Aside from the gross tonnage of the ship, two other options of "the length of the ship" and "the bunkers capacity" were also considered.[126] The IMO Legal Committee decided that "gross tonnage" was the preferred solution.[127] "Gross tonnage" is the common approach used in the IMO Conventions; hence it is more readily understood and can be easily administered by both the flag States and the port States.[128]

According to the Bunkers Convention, the obligation to take out insurance is imposed merely on the ship with a gross tonnage of above 1,000. Not all vessels are thus required to obtain insurance. A large number of small ships having a gross tonnage of less than 1,000 which may be able to cause pollution damage are therefore not covered.

Two relevant questions may be asked regarding "gross tonnage": 1) what is the relationship between gross tonnage and the bunker oil capacity? 2) What is the lowest gross tonnage at which persistent fuel oil may be used and cause substantial pollution damage?

Throughout the preparatory work of the Bunkers Convention, the insurance threshold figure was intensely discussed, since it was regarded as one of the important factors in producing administrative efficiency and determining the applicability of the Bunkers Convention. China, India, Vanuatu and Hong Kong, China, proposed that the insurance threshold should cover ships having a gross tonnage of not less than 5,000.[129] This figure was much higher than the amount written consequently in the Bunkers Convention. The reasons for this proposal were as follows:

(i) Most ships under 2,000 gross tonnage carry much less fuel oil and use only diesel oil as fuel, whereas 10,000 and 20,000 gross tonnage ships carry nearly 1,000 tons and 2,000 tons of persistent oil respectively;
(ii) The liability rules and insurance requirement in the CLCs are applicable to ships carrying more than 2,000 tons of oil as cargo. Similarly, the Bunkers Convention should only apply to cargo ships carrying more than 2,000 tons of bunker oil;

The Ratification of Maritime Conventions (2004), Vol.2, II.3.70. It was the first successful attempt to introduce a universal tonnage measurement system.

126 See IMO LEG 79/6/1.

127 See IMO LEG 82/3/2.

128 See IMO LEG 79/6/1.

129 See IMO LEG/CONF.12/7: they proposed that fishing vessels should not be considered, as their operations are not similar to those of cargo ships. Fishing ships are restricted to fishing in certain areas and their operation may vary according to the region. Other specialised ships such as refrigerator ships, dredgers and research vessels were not considered either, as the number of ships of these types is limited.

(iii) The manual on "Oil Pollution Contingency Planning" considered an oil spill from a cargo ship of 10,000 gross tons to be of no great concern and that it should be within the capability of individual facilities or of the harbour authority;

(iv) As far as the structure of the bunker is concerned, bunker fuel is distributed in four or more double-bottom tanks and a breach of all the tanks is unlikely. Therefore, not all bunker oil inside the double-bottom tanks will pollute the environment if the vessel runs aground.

The International Chamber of Shipping, in consideration of the reduction of the administrative burden on States and the industry, to some extent supported the aforesaid proposal, saying that: [130]

> "A figure which is reasonable and manageable in practice should be agreed. It was observed in the paper submitted by Hong Kong, China, to the eighty-second session of the Legal Committee[131] that most of the ships under 2,000 gross tonnage in their sample used only diesel oil as fuel. In our view, the proposed convention should target spills of bunker oil (persistent oil) and not diesel oil."

Comparing the cost of oil-removal operation in pollution cases involving vessels with compulsory insurance with cases involving vessels without compulsory insurance, the United States did not agree with the gross tonnage proposed above. It insisted that compulsory insurance should apply to vessels of 300 to 500 gross tons.[132] Australia also argued[133] that 400 gross tons, a balance between the coverage of ships and the administrative burden,[134] was preferred as it most closely reflected current practice in international conventions.[135] All these proposals were based on an idea that an extensive range of vessels should be covered by the proposed bunkers convention.

In order to help resolve the issue, the Lloyd's Register of Shipping, in a letter to the IMO Secretariat dated 31 October 2000, provided some information, based on data, about the relationship between vessels' gross tonnage and bunker fuel oil capacity.[136]

---

[130] See IMO LEG/CONF. 12/10.

[131] See IMO LEG/CONF. 82/3.

[132] See IMO LEG/CONF. 12/12.

[133] See IMO LEG/CONF. 12/6.

[134] Since the Bunkers Convention covers any type of vessel, the provision of compulsory insurance would lead to excessive administrative burden on maritime administrations. Therefore, giving exemptions to smaller ships can ease the administrative burden.

[135] Australia has adopted domestic legislation and as from 6 April 2001 requires all ships of 400 or more tons entering or leaving Australian ports to maintain specific insurance to cover the cost of a clear-up resulting from the spillage of bunker fuel (used in ships' engines) or other oil.

[136] See IMO LEG/CONF.12/4.

Eventually, at the end of day, the threshold figure of 1,000 gross tons was proposed by the Conference Chairman, *Alfred Popp Q.C.*[137] This relatively high threshold could not satisfy all delegations, because there was a concern that many vessels which may cause serious damage due to a bunker spill would escape the requirement of compulsory insurance.[138] The amount of 1,000 or higher was established probably due to the desire to arrive at a convention.

Hence, ships with less than 1,000 gross tonnage are not required to take out insurance under the Bunkers Convention. The Convention neglects, although maybe not purposely, the "small ship"[139] liability issue in the case of bunker-oil spills. Small ships can cause pollution damage which may even be quite serious. Nevertheless, the strict liability provisions of the Bunkers Convention will have an influence upon all types of vessel whether they are big or small. Although the small ships are not required to take out insurance, they can purchase insurance voluntarily.

c) "In an amount equal to the limits of liability..." - amount of limitation

There are no similar provisions regarding the limitation of liability in the Bunkers Convention as opposed to the CLCs, which provide that:

> "The owner of a ship... shall be required to maintain insurance or other financial security... in the sums fixed by applying the limits of liability prescribed in Article V, paragraph 1 to cover his liability for pollution damage under this Convention."[140]

The limitation amount under the Bunkers Convention is to be ascertained under "any applicable national or international limitation regime". Without providing a certain amount, this article is quite obscure and uncertain under the Convention itself. This problem may be solved if all Party States to the Bunkers Convention could ratify the 1996 Protocol to the 1976 LLMC.

---

137 Griggs, Patrick, 'International Convention on Civil Liability for Bunker Oil Pollution Damage', available at: <http://www.bmla.org.uk/documents/imo-bunker-convention.htm> (visited 5 April 2005).

138 See *ibid.*

139 "Small ships" are those having gross tonnage less than 1,000 and are not required to take out insurance under the Bunkers Convention.

140 See Art. VII and Art. V.1 of the 1969 CLC and Art. 6.1 of its 1992 Protocol Art. V.1 of the 1969 CLC read: "The owner of a ship shall be entitled to limit his liability under this Convention in respect of any one incident to an aggregate amount of 2,000 francs for each ton of the ship's tonnage. However, this aggregate amount shall not in any event exceed 210 million francs". This paragraph was replaced by the following text in Art. 6.1 of the 1992 Protocol: "The owner of a ship shall be entitled to limit his liability under this Convention in respect of any one incident to an aggregate amount calculated as follows: (a) 3 million units of account for a ship not exceeding 5,000 units of tonnage; (b) for a ship with a tonnage in excess thereof, for each additional unit of tonnage, 420 units of account in addition to the amount mentioned in subparagraph (a); provided, however, that this aggregate amount shall not in any event exceed 59.7 million units of account."

2. *Insurance certificate and its recognition*

As in the CLCs, there are detailed provisions in Article 7 of the Bunkers Convention regarding the requirements of compulsory insurance and the recognition of such insurance or financial security. Every ship registered in a State Party must maintain a certificate to attest that insurance or other financial security is in force in accordance with the Bunkers Convention. The certificate shall be in the form of the model set out in the Annex to the Convention and shall contain the stipulated particulars.[141] Otherwise, it will be invalid. The certificate written in or translated into English, French or Spanish,[142] shall be carried on board.[143]

Article 7 contains provisions regarding the authorisation of a State Party to institutions or organisations to issue the certificate.[144] The authorised institutions or organisations must issue the certificate in accordance with the Convention. They have the authority to withdraw the certificate if conditions have changed, and they shall report to the State on whose behalf the certificate was issued.[145] Nevertheless, the State Party shall guarantee the completeness and accuracy of the certificate and shall notify the Secretary-General of the specific conditions of the authorisation as follows: (i) the specific responsibilities and conditions of the authority delegated to an institution or organisation recognised by it; (ii) the withdrawal of such authority; and (iii) the date from which such authority or withdrawal of such authority takes effect.[146] Since the Bunkers Convention will involve all types of vessel, these provisions will undoubtedly put an enormous administrative burden on the State Parties. This aspect will be discussed in more detail in Chapter 4.

The same article contains the provision on the international recognition of the certificates. It imposes obligations and confers rights on the State Parties. On the one hand, a State Party shall accept the certificates issued or certified under the authority of other State Parties and shall regard the certificates as having the same force as if the certificates were issued or certified by itself.[147] On the other, "...A State Party may at any time request consultation with the issuing or certifying State should it believe that the insurer or guarantor named in the insurance certificate is not financially capable of meeting the obligations imposed by this Convention."[148]

During the discussion, the question also emerged whether the Convention should require a ship to carry an insurance certificate in which relevant records were maintained in an electronic format. Eventually, the Convention provides that a ship is allowed to rely on electronic certificates when entering or leaving the ports of State Parties, provided that the State Party that issues the certificate to the

[141] The Bunkers Convention, Art. 7.2 and the Annex.
[142] The Bunkers Convention, Art. 7(4).
[143] The Bunkers Convention, Art. 7(5).
[144] The Bunkers Convention, Art. 7(3).
[145] The Bunkers Convention, Art. 7(3)(c).
[146] The Bunkers Convention, Art.7(3)(b)(iii).
[147] The Bunkers Convention, Art. 7(9).
[148] The Bunkers Convention, Art.7(9).

ship has notified the Secretary-General that it maintains records of such a certificate in an electronic format.[149] But this provision only applies to State Parties.[150]

An exception to the requirement of compulsory insurance and financial security provided for in Article 7(14) is made for State-owned ships and says:

> "If insurance or other financial security is not maintained in respect of a ship owned by a State Party, the provisions of this Article relating thereto shall not be applicable to such ship..."[151]

In effect, whether or not any provisions in the Bunkers Convention are applicable to the State-owned ships on Government non-commercial service is under the discretion of the State Party.[152] Accordingly, the provisions as regards *compulsory insurance or financial security* can generally be chosen not to apply to State-owned ships. Nevertheless, these State-owned ships are required to carry a certificate issued by the appropriate authority of the State of the ship's registry, guaranteeing that the ship is owned by that State and its liability can be covered sufficiently in accordance with the Convention.[153] However, when State-owned ships are used for commercial purpose, all relevant provisions in the Bunkers Convention would be applicable to them. In this case, each State shall "waive all defences based on its status as a sovereign State".[154]

There is another exception regarding compulsory insurance for ships operated exclusively within the area of that State referred to in Article 2(a)(i) of the Convention, i.e. in the territory, including the territorial sea, of a State Party.[155] During the negotiations, the issue that arose was whether to extend this exception to a much wider area including the States' exclusive economic zones. In the end, the conference adopted a compromise that the exclusion would only apply to the territorial sea if the State itself made a pertinent declaration at the time of ratification, acceptance, approval of or accession to the Convention.[156]

3. *Direct action against the insurer*

The concept of direct action against the insurer, which was approved for the first time in the 1969 CLC, has become one of the distinct features of the insurance

149 The Bunkers Convention, Art.7(13).
150 See IMO LEG/CONF.12/13.
151 The Bunkers Convention, Art. 7(14).
152 The Bunkers Convention, Art. 4(3).
153 The Bunkers Convention, Art. 7(14).
154 The Bunkers Convention, Art. 4 (4).
155 The Bunkers Convention, Art.7(15). This article allows State Parties to declare that Article 7 does not apply to ships operating exclusively within their territory or territorial sea. However, it is important to be aware that the risk from bunker fuel is the same regardless as to whether the ship is engaged on a domestic or international voyage. It is thus important for the State Parties to take corresponding measures in this respect.
156 The Bunkers Convention, Art. 7(15).

obligation under international civil liability conventions. The purpose of direct action is to provide a better security for claimants, even in the circumstance where the shipowner is not entitled to a claim on the insurance policy due to policy defences. The exercise of the right of direct action is not limited to the claims brought against the insurer, who is bound by the compulsory insurance. The claimants may also claim directly against the liability insurer, should there be one, in the case of small vessels with a gross tonnage of less than 1,000 which is not covered by the compulsory insurance requirement.

Under the Bunkers Convention, a person claiming compensation for pollution damage may bring that claim directly against the insurer or other persons providing financial security. Meanwhile, there are certain rights granted to the insurer. First, the insurer may be entitled to invoke the defences which the shipowner would have been entitled to invoke other than bankruptcy or winding up of the shipowner, including the limitation pursuant to Article 6. Secondly, even if the shipowner is not entitled to limit his liability according to Article 6, the insurers may limit their liability to an amount calculated in accordance with Article 7(1).[157] In addition, the insurer may avoid liability if he can establish that the damage resulted from the wilful misconduct of the shipowner.[158] In this case, the assured himself will bear the entire compensation burden. There are no other defences, which may exist in normal circumstances that are available to the insurer under the Convention. Those defences apply to the person providing financial security as well.[159] The limitation of liability and the limit of insurance will be discussed in Chapter 5. More about a third party's right to direct action will be examined in Chapter 6 of this research.

4. *Time limit for bringing an action*

The Bunkers Convention entitles any pollution victim including a State to compensation. For the right to claim compensation, Article 8 of the Bunkers Convention provides:

> "Rights to compensation under this Convention shall be extinguished unless an action is brought thereunder within three years from the date when the damage occurred. However, in no case shall an action be brought more than six years from the date of the incident which caused the damage. Where the incident consists of a series of occurrences, the six-year period shall run from the date of the first such occurrence."[160]

Apparently, it contains two different time limits and one particular situation - "where the incident consists of a series of occurrences". Those two different time limits are: (1) Three years: the right to claim compensation is extinguished if an action is not brought within three years from the date *when the damage occurred.* (2) Six years: it will be possible that in some difficult cases, the three-year time

[157] The Bunkers Convention, Art.7(10).
[158] The Bunkers Convention, Art.7(10).
[159] The Bunkers Convention, Art.7(10).
[160] The Bunkers Convention, Art.8.

limit is not enough for the pollution victims to find or bring an action. So the Convention provides the victims with a six-year time limit. The right to compensation is, however, barred if the action cannot be brought after six years from *the date of the incident which caused the damage.* In particular, the said article specifies: "where the incident consists of a series of occurrences, the six-year period shall run from the date of the first such occurrence."[161] That is reasonable since one "incident" may involve a series of occurrences having the same origin.[162]

The claimants should be aware of the different terms in Article 8 such as "from the date when the damage occurred" and "from the date of the incident which caused the damage". The time limit thus starts to run at different points in time. Once pollution damage occurred to the victims, the time limit starts to run from the date when the damage occurred. It is admitted that the claim can only arise when the damage was caused or when the preventive measures were taken. It is assumed that victims would file their claim against the shipowner immediately after pollution damage occurred. However, it may occur that the spilt or discharged oil causes continuing damage over a long period of time, or long-term measures are required to take to prevent or minimise pollution damage, especially where pollution damage does not appear until some years after the incident which caused the damage.[163] In those cases, the six-year period may be applicable and the time limit starts to run from the date of the incident which caused the damage.

VI. Jurisdiction, recognition and enforcement

If pollution damage is sustained only in one country, the situation is less complicated and there is no need for any further rules to identify jurisdiction. However, if the same incident has resulted in pollution damage along the coasts of at least two countries and the courts of these countries have competent jurisdictions, claimants in each country may not be willing to seek compensation in the courts of another country. It would be more problematic if there were different rules regarding liability and its limitation.

Accordingly, Article 9(1) of the Bunkers Convention provides the rule of jurisdiction as follows:

> "Where an incident has caused pollution damage in the territory, including the territorial sea, or in an area referred to in Article 2(a)(ii) of one or more States Parties, or preventive measures have been taken to prevent or minimise pollution damage in such territory, including the territorial sea, or in such area, actions for compensation against the shipowner, insurer or other person providing security for the shipowner's liability may be brought only in the courts of any such States Parties."[164]

---

[161] The Bunkers Convention, Art.8.
[162] The Bunkers Convention, Art.1(8).
[163] Wu, Chao, *supra*, note 3, p. 72.
[164] The Bunkers Convention, Art.9(1).

The situation would not be so complex if pollution damage as well as its preventive measures occurred in the same State. In this case, the courts of the said States would have jurisdiction. However, facts of an incident always turn out to be not so simple. It is more probable that for the same incident, different claimants will bring the actions for compensation either in the court of the State or States in which the damage has been sustained, or in the courts of the State where measures to prevent or minimise pollution were undertaken.

The State Party shall ensure the competence of its jurisdiction. It shall ensure that its court has jurisdiction to consider and accept the actions for compensation under the Bunkers Convention. Reasonable notice of any such action shall be given to each defendant involved.[165]

Under Article 10, a judgment rendered by national courts following pollution incidents covered by the Bunkers Convention can be recognised and enforced by the courts of other member States. No court can challenge the decision of another court unless the judgment was obtained by fraud or the defendant was not given reasonable notice and a fair opportunity to present his or her case.[166] However, if there is more than one competent court and the viewpoints of the courts are not completely consistent, it will be difficult to ascertain how the liability can be finally decided and how to satisfy all the State Parties involved.[167] The answer has to be carefully considered. It would, however, exceed the goal of this book if more detailed research were carried out on this point.

It may occur that a shipowner purchases only one insurance policy for all liability risks incidental to his adventure. In this case, it is interesting to analyse the conflicts between the jurisdiction as regards pollution damage and the jurisdiction chosen for other types of liabilities covered by the same insurance policy. If liability issues arise, there may be different competent courts. The former jurisdiction needs to follow the jurisdiction rule under the Bunkers Convention. However, the latter still has to comply with the terms in the insurance contract. For instance, the P&I Club Rulebooks normally provide a clause which refers disputes between a member or a co-assured and the Association to the specific jurisdiction of one court or to arbitration; the applicable law is also specified.[168]

---

165 The Bunkers Convention, Art.9(2), Art. 9(3).

166 The Bunkers Convention, Art.10.

167 Wu, Chao, *supra*, note 3, p. 74, for similar issue under CLCs.

168 Gauci, Gotthard, *Oil Pollution at Sea: Civil Liability and Compensation for Damage* (1997), pp. 252-253. See also Hazelwood, Steven J., *P&I Clubs Law and Practice* (2000), p. 57: "The reason for including a jurisdiction clause in Club Rules is based upon the principle that issues involving the mutual association of members should not be solved in different ways according to the various national home states of the international body of members, but that all issues should be decided in accordance with one particular regime in order to achieve uniformity, or mutuality, or treatment throughout the international spread of membership."

VII. Other matters

Articles 11 to 19 follow nearly the format of all international conventions dealing with separate issues regarding supersession clause [169]; signature, ratification, acceptance, approval and accession[170]; States with more than one system of law[171]; entry into force[172]; denunciation[173]; revision or amendment[174]; depositary[175]; transmission to United Nations[176]; and languages[177].

Two of them require particular comments. The first relates to States with more than one system of law. This provision was inserted at the request of Hong Kong, China, since Hong Kong, China currently applies a different legal system from the mainland of China. Article 13 thus enables the State, at the time of signature, ratification, acceptance, approval or accession, to declare the Bunkers Convention as extending to all its territorial units or only to one or more of them and the State may modify the declaration by submitting another declaration at any time.[178]

The second concerns the entry into force of the Convention.[179] There was a detailed analysis in a paper[180]submitted by Australia about this issue. It contained matters such as: (a) timing of entry into force; (b) number of States required for entry into force; (c) tonnage requirement; (d) entry into force for States which accede after the Convention has entered into force. Australia came up with a proposal as follows:

(1) The present Convention shall enter into force 12 months following the date on which 12 States have either signed it without reservation as to ratification, acceptance or approval or have deposited instruments of ratification, acceptance, approval or accession with the Secretary-General.
(2) For any State which ratifies, accepts, approves or accedes to it after the conditions in paragraph 1 for entry into force have been met, the present Convention shall come into force 3 months after deposit by such State of the appropriate instrument.

It was proposed that a tonnage requirement should be carefully considered since an inappropriate result would frustrate the early entry into force of the Bunkers Convention.

---

169 The Bunkers Convention, Art. 11.
170 The Bunkers Convention, Art. 12.
171 The Bunkers Convention, Art. 13.
172 The Bunkers Convention, Art. 14.
173 The Bunkers Convention, Art. 15.
174 The Bunkers Convention, Art. 16.
175 The Bunkers Convention, Art. 17.
176 The Bunkers Convention, Art. 18.
177 The Bunkers Convention , Art. 19.
178 This provision is not exclusive to the Bunkers Convention. For example, it can be found in Article 13 of the Convention on Arrest of Ships, which was adopted in Geneva in March 1999.
179 The Bunkers Convention, Art.14.
180 See IMO LEG/CONF.12/6.

Eventually, considering the possible need for a proper and effective preparation and enforcement of insurance certificates, the Bunkers Convention adopted the proposal that it shall enter into force one year following the date on which 18 States, including 5 States, each with ships not less than 1 million gross tonnage, have ratified.[181] One may imagine the difficulties of the drafters in making such a decision. On the one hand, this unusually high number of States[182] and the gross tonnage requirement might mean that the Bunkers Convention will not come into force in the near future. Almost five years after the Bunkers Convention was adopted, there are only 9 Contracting States representing 9.07% of the world tonnage.[183] On the other hand, it can provide adequate time for the States to be advised and to finalise the administrative arrangements pursuant to the Convention.

VIII. Adopted resolutions

As mentioned earlier, the conference on the Bunkers Convention also adopted three resolutions together with the adoption of the Bunkers Convention itself. They are: (1) a resolution on limitation of liability; (2) a resolution on promoting technical co-operation and (3) a resolution on protection for persons taking measures to prevent or minimise the effects of oil pollution.

The first resolution, the resolution on limitation of liability, deals with the possibility of having uniform limitation rules governing the liability of bunker-oil pollution. It has been suggested that there exists great uncertainty as regards limitation of liability. This resolution thus urges all States that have not yet done so to ratify or accede to the Protocol of 1996 to amend the Convention on Limitation of Liability for Maritime Claims, 1976. More details will be examined in Chapter 5.

The second resolution is on the promotion of technical co-operation. The resolution urges all IMO Member States, in co-operation with the IMO, other interested States, competent international or regional organisations and industry programmes to promote and provide directly, or through the IMO, support to States that request technical assistance for: (a) assessment of the implications of ratifying, accepting, approving, or acceding to and complying with the Convention; (b) the development of national legislation to give effect to the Convention; (c) the introduction of other measures for, and the training of personnel charged with, the effective implementation and enforcement of the Convention.

181 The Bunkers Convention, Art.14(1).

182 In the 1969 CLC, the requirement in this respect is eight States; in the CLC Protocol 1992, the requirement is ten States; and even the usual entry into force requirement for a technical Convention such as MARPOL is only 15 States.

183 The relevant information is available at: <http://www.imo.org/Conventions/mainframe.asp?topic_id=247> (visited 3 January 2006). These 9 Contracting States include: Cyprus, Greece, Jamaica, Latvia, Luxembourg, Samoa, Slovenia, Spain and Tonga.

The resolution also urges all States to initiate action without waiting for the entry into force of the Bunkers Convention. Obviously, all of these aspects are very important and necessary for either the ratification or the implementation of the Bunkers Convention.

The third resolution is for the purpose of protecting persons taking measures to prevent or minimise the effects of oil pollution. The resolution urges the States, when implementing the Convention, to consider the need to introduce legal provisions to protect persons taking measures to prevent or minimise the effects of bunker-oil pollution. It recommends that persons taking reasonable measures to prevent or minimise the effects of oil pollution be exempt from liability unless the liability in question resulted from their personal act or omission, committed with the intent to cause damage, or with recklessness and with knowledge that such damage would probably result.

In addition, the third resolution recommends that all States consider the relevant provisions of the HNS Convention[184] as a model for their legislation. Due to the fact that the Bunkers Convention has removed the similar provision of "channelling of liability", pollution victims can thus claim for compensation for bunker-oil pollution damage against any person taking preventive measures. It will discourage the undertaking of preventive measures, since should pollution damage occur, the person taking preventive measures might be held liable. They may be liable under ordinary tort law to pollution victims or to the shipowner or even to both. By contrast, the relevant provision in Article 7(5) in this respect in the HNS Convention, which is in effect similar to the one contained in the CLC Protocol 1992,[185] provides:

> "Subject to paragraph 6, no claim for compensation for damage under this Convention or otherwise may be made against:
> x x xx x xx x x
> (d) any person performing salvage operations with the consent of the owner or on the instructions of a competent public authority;
> (e) any person taking preventive measures;
> (f) the servants or agents of persons mentioned in (c), (d) and (e);
> unless the damage resulted from their personal act or omission, committed with the intent to cause such damage, or recklessly and with knowledge that such damage would probably result."

It is usual to channel all liability to the shipowner by exempting other groups of persons from the liability exposure, as stated above. However, the Bunkers Convention does not contain similar provisions.

The above-mentioned resolutions cannot be expected to have the same effect as what is stipulated in the Convention, since a State can choose not to accept any or all of these resolutions.

---

[184] The HNS Convention, see *supra*, note 121.

[185] CLC Protocol 1992, Art.4(2).

D. Concluding remarks

In practice, the number of non-tanker oil spills is significantly greater than the number of tanker oil spills. Moreover, the nature of fuel oil itself makes such spills more difficult and more costly to clean up. Dealing with bunker spill from a non-tanker is difficult due to the lack of an international liability and compensation regime. The convention in this respect is thus necessary.

Commenting on the successful outcome of the conference, Mr. William A. O'Neil, the former IMO Secretary-General, said:

> "The adoption of a bunkers convention completes the task initiated by the Legal Committee when it was established by IMO more than 30 years ago – namely, the adoption of a comprehensive set of unified international rules governing the award of prompt and effective compensation to all victims of ship-sourced pollution."[186]

However, whether the measures adopted in the Bunkers Convention will be smoothly carried out and the objectives of the Bunkers Convention will be achieved remains to be seen. Some separate analysis especially regarding compulsory insurance and compensation will be made in the following chapters.

---

[186] The information is available at: <http://www.imo.org/Newsroom/contents.asp?topic_id=67&doc_id=457> (visited 24 September 2003).

Chapter 2: The Birth of Compulsory Insurance for Oil Pollution Liability

A. Introduction

Marine insurance plays a significant role in the smooth development of maritime transportation and international trade. The owner of a ship, during frequent maritime adventures, will sustain loss of or damage to his ship or incur liabilities to other third parties for damage caused by his ship. In such cases, one cannot underestimate the importance of marine insurance. It is the means "by which those who own or are interested in or responsible for maritime property seek to protect themselves in respect of loss of or damage to it and against liabilities to other parties falling upon them arising out of their ownership, interest or responsibility."[1]

Maritime transportation is a high-risk business. In particular, once a large amount of oil is spilt during any maritime transportation, the cost involved can be catastrophic not only to the business, but also to the environment. This has been recognised worldwide since *Torrey Canyon* incident,[2] which was, in terms of size and effect, one of the most significant oil spills in history, as it established and developed the international civil liability system.[3] The said incident resulted in a large number of pollution claims and it revealed that the law on liability and its limitation at the time was quite insufficient to cover massive claims. It reportedly cost the United Kingdom and France three-and-a-quarter million pounds and 41 million francs respectively.[4] A new system with adequate compensation was urgently needed to satisfy innocent pollution victims.

For the new system, the insurance arrangement was prompted by the huge and particular liabilities involved, since it was recognised that the efficient working of any new system for oil pollution liability and compensation was dependent upon insurability.[5] Accordingly, under the CLCs, the shipowner was the chosen party to be liable for pollution damage. The obligation to take out insurance was also

---

1 Hurd, Howard B., *The Law and Practice of Marine Insurance relating to Collision Damages and Other Liabilities to Third Parties* (1952), p. xxi, in: "Introduction".

2 For more detailed information regarding the international civil liability system see Chapter 1, Section A.

3 The relevant facts of the *Torrey Canyon* see Chapter 1, Section A.

4 Wu, Chao, *Pollution from the Carriage of Oil by Sea: Liability and Compensation* (1996), p. 9.

5 See Documentation C.M.I. 1968-I, p. 80.

imposed on him. Compulsory insurance or financial security requirement under the CLCs was considered to ensure the payment of prompt and effective compensation to pollution victims, which is also the main purpose under the Bunkers Convention.

The implementation of compulsory insurance is crucial. The introduction of a compulsory insurance or financial security system in international civil liability conventions could not automatically create an insurance market for it. It was the precondition that the drafters of such a convention should consider the availability of insurance from the beginning; however, it was also possible that the insurance market agreed to face up to the new challenge voluntarily.

This chapter will firstly explain the concept of compulsory insurance and its importance. In order to understand the implication of oil pollution liability for the insurance industry, it is important to have an idea of the basic framework of the marine insurance market. Following the review of the types of marine insurance and their coverage, it will help us understand the structure of the current marine insurance market and also set the stage for the discussion of P&I insurance and other types of financial security available for oil pollution liability in practice.

B. The concept of compulsory insurance

I. Development of the concept of compulsory insurance

Before 1969, insurance for any liability was an internal matter for shipowners.[6] Even at present, to insure against physical loss or damage, a shipowner can choose the insurer from the insurance market and the terms of the insurance contract. The actions of other persons do not affect his right to recover from the insurer. For third party liabilities, most shipowners join one of the mutual insurance associations comprising the International Group of Protection and Indemnity Clubs, which will be described in detail in the following sections of this chapter. It should be particularly noted that a shipowner, when entering his P&I Club, is not obliged to insure all the risks set out in the Club Rule, but may choose certain risks which the P&I Club covers, and which he perceives as most pressing.[7] Bearing in mind these advantages, a shipowner may bargain and protect his interests when entering the Club.

Although insuring liabilities was made compulsory much earlier in other branches of insurance law, for instance in insurance for motor vehicles, the shipowner's liability insurance for oil pollution damage was not made compulsory until the advent of the 1969 CLC. One of the crucial issues for urgent study after *Torrey Canyon* was whether it would be advisable to make some form of

---

6 Rosag, Erik, 'Compulsory Maritime Insurance', *Scandinavian Institute of Maritime Law Yearbook 2000*, available at: <http://folk.uio.no/erikro/WWW/corrgr/insurance/simply.pdf> (visited 24 May 2005), at 1.

7 The information is available at: <http://www.ukpandi.com/> (visited 10 April 2005).

insurance against oil pollution liability compulsory. If it was plausible, another question was whether the advantages of establishing such a system would outweigh its practical difficulties.

For the enquiry arising from the wreck of the *Torrey Canyon*, the C.M.I co-operated with the IMCO.[8] A number of reports opposed the idea of compulsory insurance of oil pollution liability[9], since the practical difficulties were likely to outweigh the advantages. The observations were as follows:

(1) Such an obligation serves the sole purpose of ensuring solvency. However, solvency matters not only for oil pollution claims but also for all other claims. Requiring security for one class of claims only means putting a particular class of claimants in a superior position over all other claimants.[10]
(2) It will impose an additional burden on shipping to deal with an extremely remote contingency.[11]
(3) The insurance cover will involve high premiums, which may prove difficult or even be impossible to pass on because freight as well as oil prices are market prices. It is impossible to predict whether the market can absorb the additional price involved in the proposed insurance scheme.[12]
(4) There is no international uniform system of insurance conditions for liability risk and no uniform opinion on the adequate minimum cover. It is felt that an international compulsory insurance scheme will be unworkable unless it is adopted universally, which is, however, quite impossible to attain.[13]

In addition, there was the worry that governments would have difficulties in checking the validity of this insurance.[14]

Some reports, on the other hand, showed the preference for a compulsory insurance system.[15] The main reason presented for this preference was that the introduction of such an insurance system would accomplish the purpose of the

8 The C.M.I. is the brief name of "Comite Maritime International"; "IMCO" is the former name of current IMO-"International Maritime Organization". They have appeared in Section B.II of Chapter 1.

9 For example, the reports submitted by the Danish Maritime Law Association, Germany Association, the Norwegian Maritime Law Association, the British maritime Law Association.

10 See Documentation C.M.I. 1968-III, The German Maritime Law Association, at 37-38.

11 See *ibid.*, Maritime Law Association of the United States, at 140.

12 See *ibid.,* The German Maritime Law Association, at 38.

13 See *ibid.*, Maritime Law Association of the United States, at.140.

14 Wu, Chao, *supra*, note 4, p. 66. The concern was focused on: "1. the high cost of compulsory insurance on a world scale covering large risks and requiring re-insurance through several insurers; 2. the lack of capacity in the insurance market and difficulties in defining conditions and premiums; 3. the difficulties for governments in checking the validity of this insurance; 4. the discrimination that could result from compulsory insurance in that damage other than pollution damage does not require such insurance."

15 See Documentation C.M.I.1968-III. For example: The Hellenic Maritime Law Association, the Japanese Maritime Law Association.

proposed convention, i.e. to ensure effective and adequate redress for victims. This implied that the liable person had the obligation to show at all times that in the event of pollution damage, adequate funds should be made available to the victims, for instance, by way of compulsory insurance or by providing other types of financial security. It was also suggested that the insurance certificate or other similar document should exist together with other ship's documents showing that the funds, though not exceeding the limit of liability, were available to meet claims. Direct recourse of the victims should also be possible, irrespective of the solvency of the assured.[16]

Meanwhile, other kinds of financial security were also recommended, since other forms of guarantee would produce the same function as compulsory insurance. The equivalent financial securities include bank guarantees, other securities furnished by third parties or personal undertakings by first-class national or international maritime shipowning companies.[17] A certificate ascertaining the financial security should likewise be added to ship's papers.

Different opinions were advanced on this issue. The question as to whether it was desirable and practicable that the liable person should be required to provide security by insurance or otherwise was not settled until the final adoption of the 1969 CLC. Compulsory insurance was also required in the CLC Protocol 1992. Together with the direct action provision,[18] the mechanism of compulsory insurance became uniform under international civil liability conventions.

II. The system of insurance

1. *Compulsory insurance as defined in international civil liability conventions*

According to the provisions in international civil liability conventions, the owner of the ship having a gross tonnage above a certain threshold shall take out insurance or have other financial security to adequately meet his potential pollution liability under victims' claims.[19] However, the conventions do not contain any definition of "compulsory insurance". The word "compulsory" in the dictionary usually means something that must be done or something required by law or rules.[20] In effect, the provisions of "compulsory insurance" in international civil liability conventions not only require the persons involved to purchase insurance policy and provide the evidence that they have done so, but also specify the insurance requirements relating to: (1) specified liabilities; (2) a certain

---

16 See *ibid.*, The Hellenic Maritime Law Association, at.80.

17 See *ibid.*, The French Maritime Law Association, the Swedish Association of International maritime law.

18 "Direct action" principle has the advantage that the claimants are not bound by the "pay to be paid" principle. This will be given detailed elaboration in Chapter 6 of this work.

19 See, for example, the 1969 CLC, Art. VII(1).

20 For instance, in: Oxford Dictionary of Current English (1990), at 235.

amount and (3) the liability insurer, who can be sued directly by claimants. The Bunkers Convention regulates compulsory insurance or other financial security in a similar way.

2. *Insurance or other financial securities*

The importance of prompt and adequate compensation for pollution claimants is underscored in international civil liability conventions. Taking into account that it might be too restrictive for the shipowners to take out liability insurance, a shipowner is also allowed to maintain other financial security to cover his liability for pollution damage under the conventions.[21] In practice, it may depend on the insurance industry or financial institutions to apply new financial tools which they have at their disposal for oil pollution liability. However, the conventions lack specific provisions in respect of a uniform interpretation of what constitutes the acceptable insurer or who can be recognised as competent providers of financial security.

Other financial security may be maintained by different methods. One method under the CLCs deemed sufficient is a guarantee from a bank.[22] Any similar financial institution is also allowed to satisfy the requirement to cover the liability for pollution damage. Above all, it is significant that the security must be financially adequate. Other types of financial security should be regulated in the same way as liability insurance and their adequacy shall also be assessed.

Although the survey of other means of financial security will not be covered by this research, it is important to emphasise two points: first, the drafters of the conventions offering options should take into account that sufficient varieties of financial security exist and are available on the market. Secondly, the availability and scope of alternative sources of financial security contribute to the smooth operation of the international civil liability system. One should not overlook their impact. To some extent, they can relieve the heavy burden on the liability insurer.

III. The need for compulsory insurance for bunker-oil pollution liability

The need to establish a compulsory insurance mechanism under the Bunkers Convention was highlighted by the occurrences of bunker-oil spills. For instance, the ITOPF[23] extended membership to include non-tanker vessels, which was due to the fact that more than 25% of oil spills involved ships other than tankers. At the 74th session of the IMO Legal Committee for a convention regulating bunker oil pollution liability, the view was expressed that it was necessary to establish a

[21] The 1969 CLC, Art. VII(1), the same provision is in the CLC Protocol 1992.
[22] The 1969 CLC, Art. VII(1), the same provision is in the CLC Protocol 1992.
[23] The ITOPF is the brief name of "International Tanker Owners Pollution Federation Limited". It has appeared in Chapter 1.

compulsory insurance mechanism for bunker-oil pollution liability.[24] In practice, the risk of claims not being paid after a bunker spill due to the absence of adequate insurance arrangements may be rare.[25] It was, nevertheless, claimed that the recovery from a bunker-oil spill incident was sometimes problematic and impossible, which was especially the case when the owner was found to have no suitable financial security to cover his liabilities. It was thus important to find a way to ensure the availability of compensation from the owner.

The proposal to establish a compulsory insurance mechanism was emphasised again in the 75th session of the IMO Legal Committee. It was fairly recognised that compensation should be ensured to meet bunker spill clear-up costs and provide damage reparations as it was further observed that:

> "...5% to 10% by tonnage of the world merchant fleet is uninsured or potentially under insured. A conservative estimate is that these ships between them could have on board at least 90,000-180,000 tonnes (tons) of bunker oils at any time. Many vessels otherwise insured may not have satisfactory liability insurance for bunker oil spills in one of the clubs of the International Group. Industry sources note that many ships may have P&I insurance but with no cover for bunker fuel pollution liability. Further, even if vessels have adequate P&I cover for bunker pollution liabilities, this may be withdrawn if Club Rules are broken."[26]

Apparently, in the absence of the compulsory insurance requirement, the liable parties are free to choose whether or not to take out insurance for any type of liabilities. Therefore, there is a danger that those ships that may cause large pollution damage cannot pay for the resulting liabilities. Comparatively, the advantages of compulsory insurance stand out. It can at least ensure compensation to pollution victims. The effect of compulsory insurance is more noticeable if the total assets of a one-ship company after an oil-spill incident are insufficient to cover pollution damage that may have been caused. Consequently, compulsory insurance was provided in the Bunkers Convention. By allowing an action to be initiated directly against the liability insurer or other persons providing financial securities, the Bunkers Convention provides the claimants with a more secure way of accessing the assets of the liable party. However, it is worth noting that the duty of compulsory insurance does not arise in all cases. As provided, the Bunkers Convention requires only ships having a gross tonnage greater than 1,000 registered in a State Party to maintain insurance.[27]

Under the compulsory insurance requirement, it appears that shipowners will be confronted with a disadvantageous situation if they are not able to purchase insurance. In practice, it is common for the insurer to utilise different insurance rates to take account of the higher risks of substandard ships. Most shipowners

---

24 See IMO LEG 74/4/1, the draft convention includes a requirement for compulsory insurance of shipowners.

25 DelaRue, Colin/Anderson, Charles B, *Shipping and the Environment* (1998), p. 269.

26 The views were expressed at the 74th session of the IMO Legal Committee, and stated in IMO LEG 75/5/1.

27 The Bunkers Convention, Art. 7(1). See Chapter 1, Section C.V.1. b).

insure for oil pollution risk through P&I Clubs. The Clubs have also established rules to motivate their member shipowners to reduce or avoid pollution incidents. For instance, it is a common practice for the Clubs to have a condition survey upon the ship's entry into the Club. The Club Rule also states that it is a condition of entry that a vessel must be and remain, throughout the period of entry, classed with a classification society approved by the managers of the Club. Where the applicant vessel is not so classed or the member fails in any way to comply with the Club Rule relating to classification, the member ceases to be insured and the period of insurance is susceptible of termination, as may be deemed appropriate by some Clubs.[28] In order to be insured, shipowners are therefore impelled to maintain their ships within certain required standards. In the long run, this is hoped to minimise pollution damage to the marine environment.[29]

Moreover, the international nature of marine transportation necessitates a uniform system of compulsory insurance. The shipowner would be exposed to different national rules in the absence of international rules. The existence of different national rules may accordingly create different schemes of insurance or financial securities. However, upon ratification of the Bunkers Convention, the shipowners will bear in mind the scope and extent of their liability and accordingly purchase sufficient insurance. However, it may meanwhile create a disadvantage to the shipowner, since his ability to trade will be dependent upon the availability of insurance.[30]

C. Possible insurers

I. The types of marine insurance and their coverage

In order to acquire a basic understanding of the marine insurance market and the necessity of P&I Clubs' insurance undertaking covering oil pollution damage, the description of the main types of marine insurance is needed, and it will be largely based on the UK London insurance market.

The category of marine insurance varies when applying different methods or in different countries. For instance, the categories of marine insurance in the US market are different from those in the London market. In the US, marine insurance includes hull policies, protection and indemnity coverage, pollution insurance and cargo insurance. Pollution insurance is a quite distinct insurance type on the American insurance market. It emerged as a separate coverage in the US partly

---

28 Hazelwood, Steven J., *P&I Clubs Law and Practice* (2000), p. 39.

29 However, since the requirement of compulsory insurance is imposed only on the ships above a certain amount of tonnage, some will not have to take out insurance. It is thus possible to find ships operating in a poor condition and with great potential to cause pollution damage.

30 Faure, Michael (ed.), *Deterrence, Insurability, and Compensation in Environmental Liability: Future Developments in the European Union* (2003), p. 186, 187.

due to the enactment of the OPA 90.[31] Pollution coverage includes removal expenses and damages. It also includes expenses incurred in abating or avoiding discharge of oil and releases of hazardous and noxious substances.[32]

The Marine Insurance Act 1906 is the act codifying the law relating to marine insurance in the U.K.[33] Anything subject to a "lawful marine adventure" may be insured. As defined by Section 3 of the Marine Insurance Act 1906:

> "(1) Subject to the provisions of this Act, every lawful marine adventure may be the subject of a contract of marine insurance.
> (2) In particular there is a marine adventure where -
> (a) Any ship, goods or other moveables are exposed to maritime perils. Such property is in this Act referred to as 'insurable property';
> (b) The earning or acquisition of any freight, passage money, commission, profit, or other pecuniary benefit, or the security for any advances, loan, or disbursements, is endangered by the exposure of insurable property to maritime perils;
> (c)Any liability to a third party may be incurred by the owner of, or other person interested in or responsible for, insurable property, by reason of maritime perils.
> 'Marine perils' means the perils consequent on, or incidental to, the navigation of the sea, that is to say, perils of the sea, war perils, pirates, rovers, thieves, captures, seizures, restraints, and detainments of princes and peoples, jettisons, barratry, and other perils, either of the like kind or which may be designated by the policy."

The subject-matter insured under the Marine Insurance Act 1906 includes both all tangible property and any intangible interest or liability of the assured. Accordingly, the subject-matter insured mainly includes: (1) ships; (2) cargo; (3) freight; (4) earnings and commissions; (5) interests in a maritime policy; (6) liability that may be incurred by the owner or charterer of the vessel; and (7) freight, demurrage and defence.[34] In theory, they can be divided into different types of insurance. Four major types, which will be discussed here, are: (1) cargo insurance covering loss of or damage to the cargo; (2) hull and machinery insurance covering loss of or damage to the ship; (3) freight insurance on the profit which can be derived by the shipowner from the commercial use of his ship; and (4) P&I insurance covering third-party liabilities.

1. Cargo insurance and its coverage

Cargo insurance is the insurance policy taken out by the insured to protect cargoes against loss or damage while they are being transported. Cargo underwriters

---

[31] 33 U.S.C.§§2701-2761 (Supp. II 1991). The OPA 90 has given a great impetus to the need for separate pollution insurance in view of the change of position of P&I Clubs regarding their willingness to provide such coverage. More details regarding "Other insurers offering OPA insurance" see Section C.II.2 of this chapter.

[32] International Encyclopaedia of Laws, Volume 3, Transport law: United States, at 221-232.

[33] The English Maritime Insurance Act 1906 is reprinted in: Rose, F.D., *Marine Insurance: Law and Practice* (2004), in Appendices, pp. 605-627.

[34] Schoenbaum, Thomas J., *Key divergences between English and American Law of Marine Insurance* (1999), p 5.

concentrate on cargo and cargo interests. An insurable interest is fundamental to all types of insurance contract. In the shipment of cargo, it attaches to anyone who (1) may benefit from the safe arrival of the cargo; (2) may suffer a financial loss as a result of the cargo being lost or damaged; (3) may incur a liability in respect of the insured goods; or (4) may suffer a loss if the goods are detained during the period of insurance. In any of those cases, the person needs to purchase cargo insurance.

The type of risks covered by cargo insurance depends on the terms of the insurance policy. In practice, the variation of the risks depends upon the incorporation of the (A), (B) or (C) clauses of the Institute Cargo Clauses (ICC).[35] The difference among them lies in the perils covered and excluded as follows: (1) the ICC (A) clauses cover all risks subject to any specific exclusion; (2) the ICC (B) clauses cover specific perils only subject to listed exclusions; (3) the ICC (C) clauses cover the perils as in the (B) clause but with some perils omitted. It is important to see from the above that even under an "all-risks" policy, as under ICC(1982)(A), the cover is subject to a list of excluded perils. The Institute Cargo Clauses are limited to the loss of or damage to the insured goods.[36] All these three sets of clauses contain cover in respect of general average contributions, salvage charges, "sue&labour" charges and liabilities incurred in connection with "both to blame" collision clauses in a contract of carriage.[37] The insurance available, however, does not cover any third-party liabilities for any resulting pollution and contamination. Therefore, if a tanker-oil spill or bunker-oil spill occurs and causes pollution damage, pollution victims cannot claim against the cargo insurer for any recovery or compensation for pollution damage based on the cargo insurance policy.

2. *Hull insurance and its coverage*

The old standard Lloyd's S.G. Policy[38] was traditionally used to cover both ship and goods. With the expansion of international trade, ships began to venture further and wider a field and the development of separate and independent insurance arrangements for ship and freight was thus necessary. This prompted the formation of the second most important subdivision of marine insurance: hull insurance. Hull insurance covers substantial loss of or damage to the insured

35 The Institute Cargo Clauses, reprinted in: Rose, F. D., *supra*, note 33, in: Appendices, pp. 649-660.

36 For more details read Hudson, N. Geoffrey (ed.), *The Institute Clauses* (1999), Part II.

37 For instance, the Institute Cargo Clauses, cl. 1.3 states that: "This insurance is extended to indemnify the assured against such proportion of liability under the contract of affreightment 'Both to Blame Collision' Clause as is in respect of a loss recoverable hereunder. In the event of any claim by shipowners under the said Clause the Assured agree to notify the Underwriters who shall have the right, at their own cost and expense, to defend the Assured against such claim."

38 The Lloyd's S. G. Policy is annexed to the Marine Insurance Act 1906. See Hill, Julian (ed.), *O'May on Marine Insurance* (1993), pp. 8-9.

vessel, including its hull, machinery, equipment, instruments and bunkers, etc., caused by the perils of the sea such as heavy weather damage, and total loss of vessel. It could also include fire, explosion, piracy, collision liability, general average contribution, salvage charges, bursting of boilers and damage caused by fixed or floating objects.[39] On the present London Market, the "Institute Time Clauses (Hulls)" is the basic policy form.[40]

Clause 7 – "Pollution Hazard" – of the Institute Time Clauses (Hulls) provides as follows:

> "Pollution Hazard
> This insurance covers loss of or damage to the vessel caused by any governmental authority acting under the powers vested in it to prevent or mitigate a pollution hazard or damage to the environment, or threat thereof, resulting directly from damage to the Vessel for which the Underwriters are liable under this insurance, provided such act of governmental authority has not resulted from want of due diligence by the Assured. Owners, or Managers of the Vessel or any of them to prevent or mitigate such hazard or damage or threat thereof. Master, Officers, Crew or Pilots not to be considered Owners within the meaning of this Clause 7 should they hold shares in the Vessel."

This Clause started to be used in 1973 and was introduced into the Institute Clauses in 1983. Its use might be linked to the intervention by the British government in the *Torrey Canyon* case.[41] Damage caused by the oil pollution incident may fall within this Clause. However, the hull insurance policy is only intended to cover "loss of or damage to the vessel". If an oil pollution incident caused loss of or damage to the ship, the insurance may cover this type of damage. Otherwise, the hull insurer will in no way provide an indemnity for the shipowner in respect of liabilities to a pollution victim.

In addition, there is a clause dealing with "3/4ths Collision Liability" following Clause 7 in the Institute Time Clause policy.[42] A collision casualty will most likely result in oil-pollution damage. However, the Clause seeks to differentiate quite clearly between the risk coverage and exclusion.[43] As a result, the insurer has to answer for liability imposed on the assured for the loss resulting from the collision of the ship, its accessories, equipment or cargo, or by a tug used by the ship. However, the insurer expressly provides in the policy that he is not liable for any sum which the assured shall pay with reference to "pollution or contamination of any real or personal property or thing whatsoever (except other vessels with

39 The Institute Time Clauses (Hulls), Clause 6.

40 The Institute Time (Hulls) Clauses 1983 was the first of the modern comprehensive forms issued by the U.K. insurance market for the insurance of ships, all previous issues of Institute Time Clauses, Hulls, having been designed for attachment to Lloyd's S.G. form. In the United States, the American Institute Hull Clauses form is in general use for hull insurance. The Institute Time Clauses (Hulls) 1/10/83 is reprinted in: Rose, F.D., *supra,* note 33, Appendices, p. 601. It will be called "ITC (Hull)" thereafter in the research.

41 Gauci, Gotthard, *Oil Pollution at Sea: Civil Liability and Compensation for Damage* (1997), p. 206.

42 Institute Time Clauses, Hulls, Clause 8: "3/4ths Collision Liability".

43 Hudson, N. G. (ed.), *supra*, note 36, p. 119.

which the insured Vessel is in collision or property on such other vessels."[44] It is apparent that the provisions are very sensitive in their coverage for pollution-damage liability. The hull insurers routinely exclude coverage as regards pollution-damage liability. Therefore, hull insurers mainly concentrate on the insurance of ships, and oil pollution victims cannot claim against them.

3. *Freight insurance and its coverage*

The third type of marine insurance concerned is freight insurance. In the context of marine insurance, "freight" means the payment made to the shipowner for the transportation of goods, either for a particular voyage or for a specified period of time. During the normal course of transportation, the freight cannot be earned if the terms outlined in the freight agreement are not complied with. For instance, if a merchant ship was contracted to carry goods from port A to port B at a particular freight rate and it deviated from the voyage route and had to discharge the goods at port C, the owner of the goods could accordingly refuse to pay the freight, since the terms of the contract had been breached. In the said case, the owner of the vessel has to incur all the costs associated with the voyage. However, if the shipowner had freight insurance in this case, the situation would be different. Freight insurance may guard against such losses, if it purported, *inter alia*, to indemnify the owner of the vessel in cases where the goods become undeliverable. However, this type of insurance cannot be used to cover pollution liability which might arise from oil-spill incidents.

4. *Protection and Indemnity Insurance*

a) Brief introduction

What is required in the Bunkers Convention is the insurance for oil pollution liability. The term "liability" means any legal liability. In general, the insurer is involved when the assured is shown to have a legal liability to others, and only insofar as the policy terms cover such liability.[45] In practice, protection and indemnity insurance (P&I insurance) is offered by shipowners' P&I Clubs. As a mutual association, the shipowners' P&I Clubs are developed to cover shipowners' third-party liabilities and expenses arising from the owning or operation of their ships.

What is the "mutual association" and how was it created? The idea of establishing a mutual association arose in the mid-nineteenth century. Shipowners at the time found themselves burdened with increasing liabilities, while their traditional hull underwriters (Lloyds) were unable or unwilling to provide cover. In order to solve the problems, groups of shipowners borrowed the idea from the

---

44 Institute Time Clauses, Hulls, Clause 8.4.5.

45 Brown, Robert H., *Introduction to Marine Insurance: Training Notes for Brokers* (1995), Section Fourteen.

earlier mutual hull-insurance association,[46] and joined in "protecting clubs" agreeing to share each other's claims. In 1855, the Britannia and the West of England Associations were established. Other similar protecting clubs were subsequently founded. Furthermore, an indemnity class was created following the awareness of implications of liabilities to cargo interests.[47] Protection and Indemnity insurance gradually came into being and the P&I Club is currently operated as a mixture of an insurance company, a law firm and a loss adjuster.[48] The main aim of this type of insurance is to protect and indemnify shipowners against losses arising from or occasioned by certain specified occurrences for which they may be liable in respect of vessels in which they are interested as owners or otherwise and which have been entered with the Club.[49]

b) The main risks covered by the P&I Club

P&I insurance is a main means whereby the shipowners protect themselves against third-party liability claims. Events that would normally be included under a hull policy are expressly excluded from P&I coverage, irrespective of whether the requisite coverage is virtually included in a hull policy.

All P&I Clubs have Club Rulebooks that include a list of risks and specify the liabilities, costs and expenses which a member may incur in a maritime adventure. Meanwhile, P&I Clubs must be flexible in adapting their coverage to changing circumstances with the expanding liabilities of the shipowners. Therefore, as years go by, more and more risks have been added to P&I insurance. One example is oil-pollution liability.

It is not necessary for a member to be covered against all risks listed in the Club's Rulebook. He often negotiates with the Club to decide upon the risks that he wants to be covered for. The main risks covered by the P&I Clubs currently include:

(1) Personal injury to or illness or loss of life of crew members;
(2) Personal injury to or loss of life of stevedores;
(3) Personal injury to or illness or loss of life of passengers and others;
(4) Loss of personal effects;
(5) Diversion of expenses;
(6) Life salvage;
(7) Collision liabilities;
   – One-fourth collision liability

46 Mutual hull underwriting associations. The hull clubs came into existence in the early part of the 18th century. Groups of shipowners got together to insure their hull risks among themselves on a mutual basis. See Coghlin, T. G., 'Protection & Indemnity Clubs', *Lloyd's Mar. & Com. L.Q.* (1984), pp. 403-416, at 403.

47 More details read Hazelwood, Steven J., *supra*, note 28, pp. 6-8.

48 Seward, Robert C., 'The Role of Protection and Indemnity (P&I) Clubs', available at:<http://www.intertanko.com/pubupload/protection%20%20indemnity%20HK%202002.pdf#search='the%20role%20of%20P&I%20club> (visited 14 April 2005).

49 Hurd, H. B., *supra*, note 1, pp. 147-148.

– Other risks excluded from the Running Down Clause
(8) Excess collision liability loss of or damage to property other than cargo;
(9) Pollution;
(10) Towage contract liabilities;
(11) Liabilities under contracts and indemnities;
(12) Wreck liabilities;
(13) Cargo liabilities;
(14) Cargo's proportion of general average or salvage;
(15) Certain expenses of salvors;
(16) Fines;
(17) Legal costs;
(18) "Omnibus" cover;
(19) Exclusion of war risks. [50]

This research will not discuss all the above-listed risks in detail. Instead, the "omnibus" cover listed above deserves more detailed discussions, since it has become one of the major advantages of P&I cover over other types of insurance obtainable on the marine insurance markets. This rule is used frequently for new risks which arise suddenly or are in some exceptional cases not within the express provisions currently in use.[51] It provides that the P&I Club shall, at its absolute discretion, have the power to decide whether to pay a member's claim in respect of losses, liabilities, costs or expenses incidental to the business of owning, operating or managing the vessels.

In addition, the insurance cover for pollution is of particular interest, since it is related to this research. It covers oil pollution and pollution by other substances. The insurance cover for oil pollution was recognised chronologically as one of the latest risks to be added to the widening spread of P&I cover.[52] A detailed account of oil pollution cover and the role of the Clubs in the case of oil pollution will be presented in this chapter.[53]

When the charterers have established "special entries" with the Club, they can also insure the risks listed above to the extent that they may incur any liabilities on fixed-premium terms. However, the charterers are a minority group and in an individual case they may have conflicting interests with the shipowners in the Club.

---

50 From UK P&I club website: <http://www.ukpandi.com/ukpandi/infopool.nsf/ HTML/ index_global2> (visited 7 August 2003). Those are the typical forms of cover in respect of members' liabilities in the P&I Club Rule-book although they may vary from Club to Club.

51 However, "...provided only that they are within the general scope of the Club cover and are not expressly excluded elsewhere within the Rules." Available at: <http://www.ukpandi.com/ukpandi/Infopool.nsf/HTML/A051F36591E6F64C8025 6E75004EBEAD?Open&Highlight=2,Omnibus%20rule> (visited 7 August 2003).

52 Hill, Christopher/Robertson, Bill/Hazelwood, Steven J., *Introduction to P&I* (1996), p. 71.

53 See Section C.II of this chapter. The role undertaken by the Club has become increasingly important due to global sensitivity on the issue of pollution.

c) Insurance contract between the shipowner and the Club

The shipowner can obtain insurance by entering a P&I Club. The shipowner will therefore become the member of the Club and thus obtain insurance cover. The normal procedure in practice is that a shipowner or a charterer seeking to enter a Club completes and signs the application form, which contains the information and particulars such as the name, type, age of the ship, class and the name of a classification society, hull insurance details and so on.[54] If the Club accepts this application form, the details of the application form will be the basis of the contract between the Club and the accepted member. After accepting the application, the Club's manager will issue a "certificate of entry" to the new member. The "certificate of entry" is not a policy. It does not even specify the terms of the cover at all times.[55] In effect, the terms of the insurance contract between the member and the Club are to be gathered not only from the application form or the "certificate of entry", but also more importantly from the Articles or Statutes and the memorandum of Association, special arrangements and by-laws as well as the Club Rules. Therefore, P&I insurance is not the same as other types of insurance, since the Club does not usually issue an insurance policy to its members. Hence, the question is whether the above arrangement between a member and his P&I Club can be regarded as a contract of marine insurance or not.

P&I insurance is recognised as a time policy in English jurisdiction.[56] The case of *the Eurysthenes*[57], which involved a P&I Club and the owners of the vessel, provided a good explanation. The English Court of Appeal was called upon to consider, *inter alia*, whether the contract between the member and its P&I Club was a "time policy" within the meaning of Section 25 of the Marine Insurance Act

54 Hazelwood, Steven J., *supra*, note 28, p. 33. According to the author, although the details required on the Entry Form will vary according to which Club and which class an applicant is seeking entry to, the following details are generally required: (i) Ship's name; (ii) Country of registry; (iii) Port of registry; (iv) Age of vessel; (v) Type of vessel; (vi) Gross Registered Tonnage; (vii) Entered tonnage; (viii) Member's name; (ix) Member's status-owner, operator, etc; (x) Member's address; (xi) Whether entered in F.D and D. Class; (xii) Class and Classification Society; (Xiii) Trade of vessel (i.e., liner, container, tramp, tanker, etc.); (xiv) Date of commencement of risk; (xv)nationality of crew; (xvi) Name and address of managers, agents, operators, brokers; (xvii) Hull insurance details; (xviii) Evidence of valid ISM certification.

55 Hill, Christopher/Robertson, Bill/ Hazelwood, Steven J., *supra*, note 52, p. 22.

56 DelaRue, Colin M./Anderson, Charles B., *supra*, note 25, p. 710: "Cover taken out during the course of the Club year will be for less than 12 months but will still normally be a time policy within the above definition."

57 [1976] 2 Lloyd's Rep. 171, it was the case between Compania Maritima San Basilio S.A. and the Oceanus Mutual Underwriting Association (Bermuda) Ltd. The latter was a P&I Club with whom the former had entered their vessel *Eurysthenes* for certain risks, at 173.

1906.[58] The judgments delivered held that the contract between the plaintiff and the defendant was a time policy in that it was a contract to insure the subject matter for a definite period of time and was therefore within the meaning of Section 25 of the Marine Insurance Act 1906.[59] Therefore, the arrangement between a member and his P&I Club could be regarded as a "time policy" within the principles of marine insurance.

Meanwhile, another issue is whether the P&I insurance policy can be regarded as a "contract of insurance" under some specific statute. One author has given a good example of whether the relationship between the Club and the member could be considered to amount to a "contract of insurance" in accordance with section 1(1) of the Third Parties (Rights Against Insurers) Act 1930 in English jurisdiction.[60] He pointed out that:

> "Doubts as to whether mutual indemnities by way of contributions from members amount to 'insurance' were dispelled in the case of *Wooding v. Monmouthshire & South Wales Mutual Indemnity Society Ltd. Etc.* and again more recently in *The 'Allobrogia'*, where Slade J. decided that although the 1930 Act made no attempt to define 'contract of insurance', the arrangement between a P&I Club and its members were 'contract of insurance' within ordinary legal terminology and within the meaning of section 1(1) of the 1930 Act."[61]

Moreover, P&I is the insurance based on the principle of indemnity. In general, indemnity insurance means that the insurer enters into a contractual relationship with the insured, whereby the insurer promises to reimburse the insured for the losses sustained within the terms of the coverage of the contract. At the same time, the insured has certain duties such as payment of premium under the contract of

---

58 Section 25 provides: "Where the contract is to insure the subject matter 'at and from', or form one place to another or others, the policy is called a 'voyage policy', and where the contract is to insure the subject matter for a definite period of time, the policy is called 'time policy'. A contract for both voyage and time may be included in the same policy."

59 See *supra*, note 57, at 173, Mr. Justice Donaldson delivered the following judgment on this issue: "... In my judgment the contract is a time policy within the meaning of the Act. It insures the vessel for a definite period of time, namely, for the balance of the current policy year when first entered and thereafter for definite periods of one year until the contract is determined under r. 17 or r. 18 (Club Rule, added by the author), ..." at 177, In addition, Lord Denning, M.R. also had the judgment that: "...I think the assurance here was a 'time policy' within s. 25 of the 1906 Act..." The other judgments delivered by the other judges were similar.

60 Third Parties (Right against Insurers) Act 1930" (the 1930 Act) in English marine insurance law, reprinted in: Rose, F.D., *supra*, note 33, in Appendices, p. 631. See Hazelwood, Steven J., *supra*, note 28, p. 52, reference is also per Viscount Cave L.C., in *Cornish Mutual Assurance Company Limited v. Commissioners of Inland Revenue* [1926] A.C. 281, at 286, see also *Thomas v. Evans (Richard) & Co.* [1927] 1 K.B.33.

61 Hazelwood, Steven J., *ibid.*, p. 311, any footnote omitted.

insurance. It is also normal that there are various exclusions, exceptions and restriction applying to this contractual relationship.[62]

One of the distinctive features of P&I insurance is the system of levying calls rather than charging premiums. Under other types of insurance contract, the premium is the consideration that the assured agrees to pay for the true performance of the insurer. The calls from the Club are described as bearing a "family likeness to premiums".[63] The P&I Club Rules normally provide that a member shall contribute by means of the advance call and the supplementary call to meeting his own losses and those of other members in the Club;[64] the contributions can also meet the expenses of management, investments and reinsurance premiums. The amount or extent of calls is negotiated between the members and the managers of the Club. It is also based upon the claims or risks each member is considered to represent.

The "Certificate of Entry" always contains the date upon which the Club's insurance cover is to commence. According to the Club Rules, the cover will usually continue until noon of the next 20th February, when the Club's policy year ends. Thereafter, the "Certificate of Entry" will usually continue from one policy year to the next. The cover will be terminated in accordance with the provisions in the Rulebooks. It will cease in certain events, including the failure of the member to pay his calls; the bankruptcy or winding up of the member; the member's ship becoming a total loss; the sale of that particular ship, in whole or in part; if the vessel ceases to be in class and so forth. The insurance may also come to an end if the member wishes to terminate the cover under his own consideration.[65]

d) The International Group of P&I Clubs

The "International Group" consists of 13 P&I Clubs. Most of the Clubs are located in England.[66] Within the "International Group", the Clubs share claims with one another and buy high levels of reinsurance on a collective basis. This enables each Club to provide a much higher level of cover than what is normally available on the commercial market. This arrangement is known as a "pooling arrangement", which is virtually a "claims-sharing agreement", whereby the claims made on one Club in excess of a certain amount are shared proportionately among all the Clubs in the pool.[67]

---

62 Alcantara Leonard F./Cox, Mary A., 'OPA 90 Certificates of Financial Responsibility', *23 J. Mar. L. & Com.* 369 (1992), p. 378.

63 Hazelwood, Steven J., *supra,* note 28, p. 121, any footnote omitted.

64 Nowadays Club managers calculate the total "premium" for the year, and then proceed to call up a proportion of total amount as an advance call, leaving the remaining to be collected by means of supplementary calls.

65 Reference from UK P&I Club's website: <http://www.ukpandi.com>, also generally from Steven J. Hazelwood, *supra*, note 28, pp. 41-45.

66 The information regarding the list of constituent Clubs is available at: <http://www.ukpandi.com/ukpandi/infopool.nsf/HTML/About_IG> (visited 31 November 2005).

67 See detailed information from Hazelwood, Steven J., *supra*, note 28, pp. 385-388.

In practice, the insurance is operated on three layers. Currently, the first layer is up to US$6 million per claim, which will be paid by each Club itself. All Clubs in the "pool" share the second layer, which is the amount from US$6 million up to US$50 million. The Claims above US$50 million are the third layer and are collectively reinsured as one contract, which is said to be the biggest liability reinsurance contract in the world. The International Group of P&I Clubs currently purchases reinsurance amounting to over US$2 billion per claim.[68] Apparently, the capacity of the International Group of P&I Clubs to obtain reinsurance and absorb catastrophic losses is huge, but they also prescribe limits to some types of risk. In particular, there is a limit of insurance cover for oil pollution liability.[69]

Summing up, the types of insurance, which include hull, cargo and freight insurance, have no or, if any, very little coverage for the shipowner's liability to third parties. The cover offered by most standard marine hull and cargo policies is generally limited to loss of or damage to the insured property. Even if the liabilities incurred by the shipowner may fall within these types of insurance, the general policy of them is to exclude insurance coverage for pollution liability risks. The above description of P&I insurance, the Clubs, what they insure and how they are organised internally and externally[70] is hoped to be of some value in understanding the Clubs' undertaking of insurance obligations under international civil liability conventions.

II. The insurers offering coverage for oil-pollution liability

There has been a massive increase in the exposure of shipowners to liability claims in respect of pollution damage caused by oil from their vessels. This is largely due to the widespread awareness of the importance of protecting the marine environment and the development of the international civil liability convention system. The following part will firstly describe the insurance practice of P&I Clubs in the cases of tanker oil spill, which can provide invaluable reference for the handling of bunker-oil pollution insurance. The external challenges from other insurance providers due to the enactment of the OPA 90 in the U.S. will be discussed below.

1. The P&I Clubs

As mentioned, shipowners' P&I Clubs are the main insurers for oil-pollution liability.[71] The following discussion regarding P&I Clubs will include three

68 See: <http://www.ukpandi.com/ukpandi/infopool.nsf/HTML/About_IG> (visited 28 June 2005).

69 See Section C.II.1.b) of this chapter.

70 For more details read Steven J. Hazelwood, *supra*, note 28, and Hill, Christopher/ Robertson, Bill/Hazelwood, Steven J., *supra*, note 52.

71 The P&I Clubs may be the Clubs within the International Group of Protection and Indemnity Clubs, or outside the International Group, since the ships may be

aspects: (1) the development of the pollution liability clause; (2) limitation of liability terms in the pollution liability clause; (3) the role of P&I Clubs in relation to oil pollution.

a) Pollution liability clause

As compared with the reality that other marine insurers offer no or extremely limited pollution-related insurance, the P&I Clubs, consistent with their objective of protecting the interests of their members, provide extensive coverage.

The *pollution liability clause* in the current Club Rule is different from what was in it around forty years ago. Research carried out by one scholar at the U.K. P&I Club offices for the purpose of ascertaining whether any changes have occurred in the drafting of *pollution liability clause* in the Club Rulebook shows that:[72]

> "In the period 1965-1973, coverage for pollution liabilities appears to have been obtainable in terms of the clause entitled 'Fixed and Floating Objects', even though there is no specific reference to pollution. The 1973 Rulebook *possibly* includes insurance relating to TOVALOP in terms of 'Liability arising under Indemnities and Contracts'. In the 1975 Rulebook, the clause dealing with 'Fixed and Floating Objects'-Clause 14- still does not make any reference to pollution. However, *the wording of the same Clause 14 in the 1976 Rulebook contains a significant change*; paragraph (iv) thereof specifically states:
>
> The Association's liability for claims arising from legal liabilities for oil pollution shall be limited to such a sum as the Directors may from time determine:
>
> *Note*: The limit under this proviso (iv) as at 20 February 1976 is US $30 million each vessel each accident or occurrence.
> There is wording, similar to the above paragraph, contained in the 1980 Rulebook; however, the limit is there raised to $200 million. The 1981 Rulebook sees a further change in wording; there is a specific 'pollution clause', namely section 12. In the 1982 Rulebook, the ceiling of liability is increased to$300 million."[73]

Apparently, there have been changes in the *pollution liability clause* especially as regards oil-pollution liability. It was in 1975 that the 1969 CLC came into force. P&I Clubs immediately responded to the insurance requirement in the 1969 CLC.

---

insured by another insurance company or association. As pointed out by Abecassis, David W. (ed.), *Oil Pollution from Ships* (1985), p. 224: "In practice, almost all tankers are insured for liabilities under the Convention either with a Protection and Indemnity Club which is a member of the International Group of Protection and Indemnity Clubs, or with the International Tanker Indemnity Association of Hamilton Bermuda. These insurance companies, each of which functions on the mutual principle, share amongst themselves their oil pollution liabilities to an extent which varies from time to time, and then re-insure such liabilities on the open market."

72 Gauci, Gotthard, *supra*, note 41, pp. 210-212.

73 See *ibid.*

The 1976 Rulebook of the U.K. P&I Club firstly and expressly provided cover for the liability for oil pollution with a specific limitation.

The Club's cover in respect of pollution has now extended to pollution by substances other than oil. It covers "any liabilities, losses, damages, costs and expenses in so far as they are caused by or incurred by reason of the discharge or escape or threat of such discharge or escape of oil or any other substance from the entered vessel."[74] More specifically, for instance, the pollution cover in the current Rulebook[75] of the North of England Protection and Indemnity Association Limited is categorised as follows: (a) Damages or compensation paid; (b) Prevention and clean-up expenses including liability for losses or damages caused by any measure so taken; (c) Agreements and Contracts[76]; (d) The costs of liabilities incurred as a result of compliance with the Government or Authority Order and Directions for the purpose of preventing or reducing pollution or the risk of pollution; (e) Salvors' Expenses or Special Compensation; (f) Fines, cover for which is sometimes specifically mentioned under the pollution coverage.[77]

b) Limitation of liability terms

The principle to limit the Clubs' exposure to oil-pollution liability was established from the first *Pollution Liability Clause.* Meanwhile, the limited amount is adjusted as the Directors of the Clubs may from time to time consider and determine.

Before the policy year 2000, the limit offered by P&I Clubs in the International Group was US$500 million for each occurrence for each vessel in the case of oil pollution.[78] P&I Clubs also made available to shipowners another US$200 million in oil-pollution coverage from the commercial market. This limitation was a corresponding reflection of the market capacity for reinsurance.[79] Therefore, many vessels could have oil-pollution coverage of up to US$700 million.

Since the policy year 2000, the Clubs in the International Group have increased the limit to US$1,000 million (US$1 billion) for each accident or occurrence in respect of each ship entered by or on behalf of an owner not being a charterer other than a demise or bareboat charterer. This was in accordance with the

---

74 Hill, Christopher/Robertson, Bill/Hazelwood, Steven J., *supra*, note 52, p. 71.

75 P&I Rules 2003/2004 of the North of England Protection and Indemnity Association Limited.

76 The Club Rules may cautiously provide that such contractual liabilities are only covered provided such agreement has been approved by the Managers and only upon an additional premium.

77 In addition, the TOVALOP liabilities, which are usually provided as "voluntary payments or operations undertaken by the member as a party to any agreement in relation to pollution approved by the directors" were ended in 1997 with the termination of this voluntary agreement.

78 For charterers' insurance, it was US$300 million.

79 Hill, Christopher/Robertson, Bill/Hazelwood, Steven J., *supra*, note 52, p. 50; Hazelwood, Steven J., *supra*, note 28, p. 227; Gauci, Gotthard, *supra*, note 41, p. 220.

decision of the International Group to increase reinsurance protection for oil pollution.[80]

c) The role of the P&I Club with regard to an oil-pollution incident

aa) Oil-pollution liability insurer

P&I Clubs can insure shipowners' oil pollution in accordance with the CLCs. Basically, the Clubs are required to step into the shoes of shipowners and be subject to direct action by pollution claimants rather than merely act as an indemnifier. The Clubs must also accept that only the defences provided in the CLCs can be invoked to avoid their liabilities. At the same time, the Clubs must provide the necessary evidence that the insurance cover for oil-pollution liability is in place as required, which enables the certificate to be issued by the relevant governmental authorities in accordance with the CLCs.

It has to be mentioned that P&I Clubs have a general policy of not providing anticipatory guarantees.[81] This position was shown by the International Group's reaction to the relevant provisions in the OPA 90. The U.S. did not ratify the CLCs and enacted the OPA 90, in which it requires the liable parties to be backed up by some form of financial security.[82] This virtually requires the financial guarantor to undertake potentially unlimited pollution liability. P&I Clubs refused to act as the guarantors to assist shipowners to obtain the certificate of financial responsibility under the OPA 90.[83]

Broadly speaking, the current Clubs' coverage in relation to pollution may also include any liabilities or expenses arising from bunker-oil spills.[84] The taking out of insurance against bunker-oil pollution liability is not compulsory. However, once the Bunkers Convention enters into force, shipowners in the Clubs who have the ship with a gross tonnage greater than 1,000 in a State Party to this Convention will be obliged to take out insurance as stipulated in Article 7(1) of the Bunkers Convention. Although the new compulsory insurance for bunker-oil pollution follows the same pattern as what is in force on tankers, it has its characteristics and difficulties. The Clubs will probably be involved in more intensive work on carrying out this insurance arrangement for all types of ships other than tankers.

In addition, shipowners may incur oil-pollution liability not only under international legal regime, but also under different national regimes. Oil-pollution

---

80 See: <http://www.ukpandi.com/ukpandi/Infopool.nsf/HTML/E8EBE7157D236C5080256DB30055E4BD?Open&Highlight=2,oil%20pollution,%202000%20policy%20year> (visited 26 May 2005).

81 Alcantara, Leonard F./Cox, Mary A., *supra*, note 62, p. 378, F.N.20, cited from Coghlin, Terence G., 'The Impact on the Country of Proposed Rules for Evidence of Financial Responsibility Required by the Oil Pollution Act of 1990': Hearing Before the Subcomm, on Coast Guard and Navigation of the House Comm. On Merchant Marine and Fisheries, 102nd Cong., 1 st Sess.(1991).

82 33 U.S.C.§ 2716.

83 More see Section C.II.2 in this chapter.

84 See Section C.II.c). aa) in this chapter.

incidents can take place in countries which are not State Parties to the CLCs or the Bunkers Convention. In those incidents, if the oil-pollution liability has been written into the Clubs' "Certificate of Entry" or endorsement slips, the Clubs shall cover it.[85] However, the Clubs often recommend their members to put the relevant clause into the Charter Party, indicating that in return for the owner warranting that he has a CLC certificate (and those required under U.S. law), the owner will have no further responsibility for establishing or maintaining any additional financial security for any particular national regime. Should any loss arise from the owner not having such additional security the owner is to be indemnified by the charterer.[86]

bb) Measures for an oil-pollution incident taken by the Clubs

(1) Clean-up or salvage operation

The Clubs need to be capable of handling a casualty for their members on any scale involving a tanker or any other types of ships in any place in the world. The Clubs usually do not undertake clean-up or salvage operation by themselves in the immediate aftermath of a casualty. They are in a position first to find appropriate advisers, surveyors and contractors capable of performing those tasks, as there are varying situations.[87] In many cases, there is little hope that the shipowner[88] or his Club can undertake preventive measures, and the authorities of the States whose coastlines are contaminated by the incident should immediately decide to assume complete control over these operations. The shipowner and his Club will then incur a financial burden by having to reimburse the costs incurred by the authorities concerned. In some other cases, if the shipowner or the oil company concerned has a highly developed emergency response plan, the Club will act as a "communicator" to interact with the interests involved and provide advice and

---

85 The risks against which the ship has been entered are normally stated in the Certificate of Entry. And during the period of membership, the Club and the owner of any entered vessel agree to vary the risks against which the vessel is covered. More details read Hazelwood, Steven J., *supra*, note 28, pp. 41-45.

86 Hazelwood, Steven J., *supra*, note 28, Appendices, pp. 454-455. Financial Responsibility for Tankers Charterparty Clause: for instance, "(2)(b) Charterers shall indemnify owners and hold them harmless in respect of any loss, damage, liability or expense (including but not limited to the cost...) whatsoever and howsoever arising which owners may sustain by reason of any requirement to establish or maintain financial security or responsibility in order to enter, remain in or leave any port, place or waters, other than to the extent provided in paragraph (1) hereof." There is a similar provision in the Financial Responsibilities for Oil Pollution for Non-Tankers Charterparty Clause. This clause is designed mainly to respond to the relevant legislation in the U.S. It might accordingly be supposed that the similar clause will apply to bunker-oil spills.

87 Seward, Robert C., *supra*, note 48, at 4.

88 The Club Rules generally contain a provision to require its member to take all steps that may be reasonable for the purpose of avoiding or minimising any expense or liability.

comments. However, if the shipowner does not have the strong manpower or resources to deal with the incident, the Club will be expected to play a main role in organising the response on behalf of the shipowner. The Club will take appropriate measures to prevent or minimise pollution resulting from the casualty and the cleaning-up of any contamination that actually occurred. In this respect, it is worth noting that the ITOPF has unrivalled expertise in undertaking preventive measures for pollution accidents.[89]

(2) Source of funding

Some shipowners may, however, need financial support to fight against a major oil pollution incident, because in some cases shipowners are not in a good enough financial situation to carry out cleaning-up measures unless the funds can be made available to them in advance. The oil companies are not willing to pay out of their own pockets for too long a time either. Therefore, the Club in such cases will be expected to be the source of funding for those preventive measures. In such a circumstance, it is important for the parties to agree on the expenditure in advance, so that any dispute can be avoided.[90]

The Club Rules generally contain a provision as to the costs or expense which the Club will cover. The costs or expenses usually include "the costs of any measures reasonably taken after the discharge or escape of oil for the purpose of avoiding or minimising any resulting loss, damage or contamination or cleaning up any resulting pollution, together with any liability for any loss of or damage to property caused by any measures so taken."[91] They also include the costs of any measures reasonably taken to prevent an imminent danger of the discharge or escape from the entered ship of oil or any other substance which may cause pollution.[92]

(3) Providing legal advice

The counselling and advice service is an important secondary function of P&I insurance.[93] If an oil-pollution incident occurs, the Club will apply the provision of "appointment of lawyers and experts" in the Rulebook and follow the rules of the Club's claim-handling facilities. The handling of oil-pollution liability may be relatively complicated and is generally divided into two integrated parts: 1) the overall issues of liability and 2) the compensation and analysis of the details of

---

89 The ITOPF has attended to 450 spills in 70 countries since 1971, including 150 bunker spills from non-tankers.

90 Seward, Robert C., *supra*, note 48, at 4.

91 For instance, a similar provision is in the Rule Book of North of England (2004/2005): available at: <http://195.173.17.24/pdf/PandI_Rules_2005.pdf> (visited 25 May 2005), at 39.

92 The UK Club's Rules 2005, available at: <http://www.ukpandi.com/ukpandi/resource.nsf/Files/Rules05_02/$FILE/Rules05_02.pdf> (visited 26 May 2005), at 7.

93 Hazelwood, Steven J., *supra*, note 28, p. 273. Claims-handling service is the first function of P&I Clubs.

particular claims.[94] In both parts, the Club is expected to assist the shipowner in order to investigate, provide the shipowner with advice and deal with any matter that may result in loss, damage, expense or liability in respect of which the member is or may be burdened.

d) Interaction with other international organisations

Unavoidably, P&I Clubs have contact with some international organisations when they share a common purpose of preventing oil pollution or protecting the marine environment. The main international organisations examined here are the International Maritime Organization (IMO), the International Oil Pollution Compensation Fund (IOPCF) and the International Tanker Owners Federation (ITOPF).

aa) The legal framework of the IMO conventions

The International Maritime Organization (IMO), as a United Nation's specialised agency, is responsible for improving maritime safety and preventing pollution from ships. The IMO has adopted a series of conventions and other instruments, including further amendments, through additional protocols, in order to formulate and develop a set of international standards for preventing pollution and providing compensation. Both the CLCs and the Bunkers Convention were adopted under IMO's auspices. Those conventions lay down the basic legal framework for related civil-liability issues.

In particular, the International Group of P&I Clubs has been granted a consultative status by the IMO since 1979. It actively took part in developing a framework for the oil-pollution compensation system. For instance, it submitted some quite detailed proposals in order to contribute to the consideration and adoption of a more practical and effective convention for bunker-oil pollution liability.[95]

Therefore, on the one hand, the IMO provides a legal framework for shipowners' liability, while the International Group of P&I Clubs positively take part in the process of preparation and negotiation; on the other hand, P&I Clubs insure the liabilities that are specified in the conventions which are adopted by the IMO.

---

94 Seward, Robert C., *supra*, note 48, at 4.

95 For example: see IMO LEG/CONF.12/9, IMO LEG 80/4/2 and IMO LEG 81/4/2. It also submitted a proposal to express the concerns of liability insurers in connection with provisions in the draft text of the bunkers convention as regards direct action, channelling, responder immunity and the definition of damage, and the proposals intended to make these aspects of the draft text workable in practice: IMO LEG 81/WP.2.

bb) Cooperation with the IOPC Fund

The IOPC Fund, as an intergovernmental organisation, was established in October 1978 and currently operates under the 1992 Fund Convention.[96] It aims at providing a tier of complementary or supplementary compensation to pollution victims.

The IOPC Fund works closely with P&I Clubs in providing compensation for oil pollution damage resulting from oil spills from tankers. The International Group of P&I Clubs has an observer status in both the 1971 Fund and the 1992 Fund. The cooperation between the 1992 Fund and the P&I Clubs in respect of the handling of incidents is embodied in a Memorandum of Understanding, which was signed in 1980 by the 1971 Fund and the International Group of P&I Clubs. This Memorandum was extended in 1996 to apply to the 1992 Fund as well.

In most large tanker-oil-spill incidents, the P&I Clubs work closely with the IOPC Fund. They may have different interests. For instance, the IOPC Fund will only become involved at a higher level for liability claims. Meanwhile, it may happen that the Fund has to commence litigation against the Club concerned. However, they cooperate with each other and share common interests in many aspects. They usually cooperate from the outset in investigating, monitoring clean-up operations, assessing and settling all claims that arise in a major oil-pollution incident, so as to ensure that smooth solutions are adopted for recovering expenditure at different levels.[97]

The Fund has recognised the Clubs' primary responsibility for handling claims against their members; at the same time, the Clubs consult the Fund concerning such claims, which would probably involve claims being made against the Fund. The Fund and the Club exchange views concerning the incident and ascertain the interpretation of relevant aspects. The many years of claim-handling experience show that the cooperation between P&I Clubs and the IOPC Fund is effective.

cc) Technical assistance from the ITOPF

The ITOPF provides a broad range of technical services in the field of oil pollution anywhere in the world. They are always on call at the request of shipowners, shipowners' P&I Clubs and any member of the ITOPF.[98] ITOPF employees have extensive, first-hand practical experience of combating marine oil spills, since they have been asked to be on site for hundreds of incidents in different countries. It is expected that the ITOPF's advice and expertise are closest

---

96 The IOPC Fund was originally operated under two international Conventions: the 1969 CLC and 1971 Fund Convention. These two conventions were later amended in 1992 by two protocols. See Chapter 1, Section A.II.

97 See Seward, Robert C., *supra*, note 48, at 3: "In large oil spills where the IOPCF may be involved the P&I Club will liaise with the IOPCF, both so as to keep them informed and so as to ensure that there is a smooth flow of funding without different views being taken on the recoverability of expenditure at different levels."

98 See more information available at: <http://www.itopf.com/>.

to international standards, such as in undertaking, as effectively as possible, clean-up measures with the minimum of damage and adopting procedures leading to acceptable compensation claims.

The ITOPF's data show that it has often been called upon to assist in cases of bunker spills from non-tankers as well.[99] At ITOPF's Annual General Meeting on 5 November 1998, in consideration of the unfair distribution of the costs incurred to maintain the organisation's expertise and technical service among its members, i.e. tanker owners and non-tanker owners, the ITOPF agreed to a special resolution to amend the organisation's Memorandum and Articles of Association, which permits non-tanker owners to become Associates of ITOPF effectively from 20 February 1999.[100] This shows that pollution damage caused by non-tankers is more and more a matter of a concern for governments and relevant international organisations.

2. *Other insurers offering OPA insurance*

The P&I Clubs offer coverage to most oil-tanker owners on an indemnity basis under the CLCs. The Clubs, as the insurers of oil-spill liability, have several advantages. Two main advantages are as follows: first, the P&I Clubs specialise in insuring marine liability and have concrete understanding of the risks. They especially have valuable insurance experience in pollution liability. Secondly, the Clubs are non-profit associations. They are able to provide insurance cover at rates that are very cost-effective. Even more distinctively, the Clubs have experience in negotiating relatively favourable reinsurance terms for their members.

In response to the enactment of the OPA 90, the International Group of P&I Clubs quickly expressed their unwillingness to act as guarantors[101] according to the financial responsibility provisions.[102] To understand the unwillingness of P&I

---

99 Although the majority of the 400 incidents attended on-site by ITOPF staffs since the mid-1970s have involved tankers, the Clubs have often called upon ITOPF to assist in cases of bunker spills from non-tankers. Whereas tanker owners who are members of the ITOPF pay dues and receive technical assistance following a spill on an expenses only basis, non-tanker owners have been charged a daily fee for such assistance. Over the last ten years, about 25% of all oil spills attended by ITOPF have been bunker spills from non-tankers. Information is available at: <http://www.american-club.com/circulars/cir24-98.htm> (visited 26 June 2005).

100 See *ibid.*

101 The OPA 90, §§ 2701(13): "'guarantor' means any person, other than the responsible party, who provides evidence of financial responsibility for a responsible party under this Act;..."

102 The International Group of P&I Clubs stated: "The Clubs are not willing to act as guarantors under OPA 90. While, as always has been the case, they are prepared to live up their obligation to pay claims pursuant to the terms of their contracts of insurance, they are not prepared to offer themselves to the US Courts as guarantors directly responsible for all claims made as a consequence of oil spills of their shipowner Members, thus in effect becoming the property insurers for claimants under OPA 90 rather than the liability insurers of shipowners." In: Certificate of Financial Responsibility under the Oil Pollution Act, 1991: Hearings on the Impact

Clubs to certify financial responsibilities under the OPA, the following part will firstly provide relevant background information on the OPA 90 and its financial responsibility requirement. Later it will describe the emergence of some specialist insurers offering OPA cover for shipowners.

a) Financial responsibility requirement in the OPA 90

The U.S. did not ratify the international civil liability conventions; on the contrary, they passed the OPA 90,[103] which was the most radical attempt made by a single nation to deal with oil spills. The OPA 90 consolidates various federal liability provisions into one statute regarding oil-spill liability, response and compensation. It has established a comprehensive system without preempting the establishment of state liability legislation or the implementation of any international oil-spill convention.[104]

One of the basic requirements under the OPA 90 is that the responsible party for any vessel over 300 gross tons shall establish and maintain the evidence of financial responsibility sufficiently to meet the maximum amount of liability to which he is subject under the OPA 90.[105] Vessels entering US waters must possess the documentation guaranteeing the maximum liability to which the responsible party could be subject.[106] This financial requirement in the OPA 90 appears to be in line with the international regime reflected in the CLCs. However, when it is considered in conjunction with other sections in the OPA 90, it shows that "...shipowners and their insurers become liable for damages far beyond any present day damage assessments."[107]

By virtue of Section 2716(a), the evidence of financial responsibility shall be sufficient to meet the maximum amount of liability subject to the limits of liability section in this Act.[108] Although the OPA 90 sets a limit on liability of a responsible party, there is in effect no maximum liability for the responsible party. A

on the Country of Proposed rules for Evidence of Financial Responsibility Required by the Oil Pollution Act of 1990 before the subcomm. On Coastal Guard and Navigation of the House Comm. On Merchant Marine and Fisheries, 102d Cong., 1 st Sess.7(1991), at 88 (statement of Terence G. Goghlin).

[103] See 33 U.S. C. §§ 2701 *et seq.*

[104] See OPA 90, §§ 2718.

[105] See OPA 90, §§ 2716.

[106] The OPA 90, §§ 2716(e): "Financial responsibility under this section may be established by any one, or by any combination, of the following methods which the Secretary (in the case of a vessel) or the President (in the case of a facility) determined to be acceptable: evidence of insurance, surety bond, guarantee, letter of credit, qualification as a self-insurance, or other evidence of financial responsibility. Any bond filed shall be issued by a bonding company authorized to do business in the United States..."

[107] Mitchell, John M., 'The United States Coastal Guard's Proposed Regulation of Certificates of Financial Responsibility under the Oil Pollution Act of 1990: Fostering a Continuing Market of Insurance for shipowners?' 7 *Admin. L. J. Am. U.*121, at 124.

[108] See the OPA 90, §§ 2716(a).

minimum liability is US$2 million for a tank vessel of 3,000 gross tons or less, and US$10 million for a tank vessel of more than 3,000 gross tons. For any other vessel, the minimum liability amount is US$500,000. Accordingly, the big difference between the OPA and the international regime is that there is no maximum liability under the OPA.[109] The limit of liability will increase according to the size of the vessel concerned.

In addition, it is easier to deprive the shipowner of his right to limit liability under the OPA 90 than that is under the CLCs. Under the OPA 90, liability cannot be limited if the incident was proximately caused by: "(A) gross negligence or willful misconduct of, or (B) the violation of an applicable Federal safety, construction, or operating regulation by, the responsible party, an agent or employee of the responsible party, or a person acting pursuant to a contractual relationship with the responsible party (except where the sole contractual arrangement arises in connection with carriage by a common carrier by rail)."[110] Mainly, there are two concerns for shipowners and their liability insurers: first, they fear that it is easy for claimants to meet the gross negligence standard because of the "unknown...nature of the test for 'gross negligence' at the shipboard level".[111] Secondly, under the above exception (B), it is also easy to lose the right to limit liability since "...even assuming that an owner can be aware of all applicable Federal safety, construction or operating regulations-a tall order for anyone especially a foreign shipowner-his right to limit can be placed in jeopardy by any

---

[109] Wu, Chao, *Pollution from the Carriage of Oil by Sea: Liability and Compensation* (1996), p. 241.

[110] In the OPA 90, §§ 2704 states:
"(c) Exceptions
(1) Acts of responsible party
Subsection (a) of this section does not apply if the incident was proximately caused by-
(A) gross negligence or willful conduct of, or
(B) the violation of an applicable Federal safety, construction, or operating regulations by,
the responsible party, an agent or employee of the responsible party, or a person acting pursuant to a contractual relationship with the responsible party (except where the sole contractual arrangement arises in connection with carriage by a common carrier by rail.)
(2) Failure or refusal of responsible party
Subsection (a) of this section does not apply if the responsible party fails to refuse-
(A) to report the incident as required by law and the responsible party knows or has reason to know of the incident;
(B) to provide all reasonable cooperation and assistance requested by a responsible official in connection with removal activities; or
(C) without sufficient cause, to comply with an order issued under subsection (c) or (e) of section 1321 of this title or the Intervention on the High Seas Act (33 U.S.C. 1471 et seq.)."

[111] Hutchinson, James A., 'Financial Responsibility Provisions: are They Sinking the U.S. Maritime Trade?' *24 Law & Pol'y Int'L Bus.* 223, at 237, 238, any footnote omitted.

of his employees or any independent contractor who has worked on the vessel."[112] It must be more difficult for a foreign-based company that is working under a different set of national safety standards to comply with every technical Coastal Guard safety or construction regulation as the OPA requires.[113] Therefore, the conditions upon which the responsible party can be deprived of the right to limit his liability are relatively easy to meet; it may result in unlimited liability in the majority of cases. In addition, the OPA does not preempt the individual U.S. states to enact their own legislation imposing higher limits of liability. This may also increase the possibility of enacting unlimited liability.

Other Sections of the OPA 90, such as Section 2703 "Defenses to Liability" which only allows very limited defences,[114] may furthermore extend the scope or extent of oil-pollution liability and related financial responsibility. The OPA 90 also enlarges the scope of compensatory damage that allows for recovery of monetary damages for the destruction of natural resources.[115]

b) The concerns of the P&I Clubs to be the guarantors

In general, as described, the Clubs provide shipowners with indemnity for pollution liability worldwide; this includes most of the oil tankers and dry cargo ships trading to the United States. However, P&I Clubs feared that potential unlimited liability under the OPA 90 could result in an unquantifiable and consequently uninsurable risk. They thus refused to be the guarantors for financial responsibility under the OPA 90. As analysed in-depth by one author, the reasons were both political and commercial:

> "...Politically the clubs feared that the effect of issuing certificates to meet specific US limits could in the future result in regional or national demands elsewhere. Commercially, the clubs were not prepared to act as guarantors under the Act and they did not want to be vulnerable to direct action from courts for damage caused by their members."[116]

The OPA 90 contains the right of direct action against any guarantor, defined as anyone who provides evidence of financial responsibility for a responsible party.[117] The guarantor is liable, jointly and severally, with the responsible party in the event of a spill. In addition, the guarantor may only invoke rights and defences that would be available to the responsible party, and the defence that the incident

112 Clark, A. F. Bessemer, 'The U.S. Oil Pollution Act of 1990', *Lloyd's Mar. & Com. L.Q.* (1991), pp. 247-256, at 250.

113 Hutchinson, James A., *supra*, note 111, at 238, footnote omitted.

114 See the OPA 90, §§ 2703. A party responsible is not liable for removal costs and damages if the party responsible establishes by a preponderance of the evidence that the discharge or substantial threat of a discharge of oil and the resulting damages or removal costs were caused solely by: (a) an act of God; (b) an act of war; (c) an act or omission of a third party.

115 See the OPA 90, §§ 2706.

116 Özçayir, Z. Oya, *Liability for Oil Pollution and Collisions* (1998), p. 286.

117 The OPA 90, §§ 2701(13).

was caused by the wilful misconduct of the responsible party. No other policy defences are available.[118]

Reading the sections in the OPA 90, the International Group was aware that: "...once exposed to direct action in the US courts as a guarantor, there are real dangers of those courts giving unexpected interpretations both to their obligations as guarantors and their obligations as insurers under the terms of an insurance contract which is, in nearly all cases, subject to a foreign law. This risk is even more significant where acceptance of the status as guarantor would have effectively turned the P&I Clubs into direct insurers of the property and environmental rights of every US citizen affected by oil spill..."[119]

The P&I Clubs paid great attention to the differences between being a guarantor and being an insurer. They insisted on maintaining the identity of an insurer rather than that of a guarantor, although, in this author's opinion, the Clubs may simultaneously be regarded as the insurer and guarantor of financial responsibility under the CLCs. The International Group maintains its firm position on not being the "guarantor" under the OPA 90. Nevertheless, the Clubs continue to provide their members with oil-spill insurance up to a certain limit.[120]

c) Some alternative ways of meeting financial responsibility requirements

The OPA 90 provides for various methods of establishing financial responsibility. They are: evidence of insurance, surety bond, guarantee, letter of credit qualification as a self-insurer, or other evidence of financial responsibility.[121] At the beginning, it was considered that any vessel in one of the Clubs with full oil-pollution cover could meet the financial responsibility requirement under the OPA 90. As pointed out by one author, several U.S. states such as California, Washington, Florida and Virginia accepted entry in a P&I Club as sufficient proof

---

[118] The OPA 90, §§ 2716(f).

[119] Readman, Luke, 'Certificate of Financial Responsibility under OPA 90: the Catalyst for an Insurance Crisis', *Int. I. L. R.*, 1994, 2(10), 361-364, at 363.

[120] See *ibid.*, p. 361. See also Hazelwood, Steven J., *supra*, note 28, p. 226: "The reaction of the clubs was to introduce an Exclusion Clause for all tankers trading to the United States which excluded all claims in respect of oil pollution in U.S. waters. Any tanker owner trading to the U.S. who wants P. & I. cover has to apply for deletion of the Exclusion Clause to reinstate his P&I Cover. In order to do so, he is required to submit accurate quarterly declarations and reports to the club of all voyages to or from the U.S.A. This will reinstate P. & I. cover provided that additional premium is paid in respect of voyage to or from the U.S.A. In this way, club cover is available for oil pollution in relation to liabilities, costs, expenses and fines arising out of an incident to which OPA 90 is applicable. Failure by a member to make a declaration in time results in excluding claims arising out of an incident to which OPA 90 applies. Failure to make an accurate declaration or provide an accurate report can result in termination of entry of particular vessel concerned."

[121] See OPA 90, §§ 2716.

of financial responsibility and even without requiring the Clubs to submit to direct action.[122]

It was stipulated in the OPA 90 that any regulation relating to financial responsibility remained in force and effect unless and until it was "superseded by a new regulation issued under this section".[123] It is thus worthwhile to recall that, in July 1994, the U.S. Coastal Guard published an interim rule implementing the vessel financial responsibility provisions of the OPA and CERCLAR.[124] This rule became the final rule in March 1996.[125]

Several alternative means emerged after the promulgation of the Coastal Guard's interim and final rule on financial responsibility, two of which were the Shoreline Mutual and the First Line programme of Stockton Re. They started to offer guarantees acceptable to the United States Coast Guard for the issuance of certificates for tanker owners.[126] Both Shoreline and First Line came from the heart of the marine insurance industry.

The Shoreline was designed by a former P&I manager, a retired shipowner and two marine insurance brokers. It had applied for and was granted full approval as a guarantor for financial responsibility under the vessel financial responsibility regulations published on July 1, 1994. The Shoreline simply covers any liability for oil or other pollution under the terms of its guarantee to the U.S. government which cannot be met by shipowners' P&I Clubs. It is a traditional, industry-standard P&I Club, mutual and non-profit-making, and controlled by a shipowners' committee, just like the International Group of P&I Clubs.

In the same year 1994 but around 4 months later, the U.S. Coast Guard's National Pollution Fund Center announced that it had approved a Certificate of Financial Responsibility (COFR) guaranty programme commonly referred to as "First Line" offered by Stockton Reinsurance Limited (Stockton Re), a Bermuda insurance company which had applied for and was granted approval as an "other evidence" provider of financial responsibility under the new financial responsibility regulations. Essentially, the "First Line" provides a credit-risk insurance covering any liability arising under a COFR guarantee which cannot be recovered from the shipowner concerned.

According to the U.S. National Pollution Fund Center, the insurers providing evidence of financial responsibility for tanker vessels (by tonnage) currently include: Sigco, Shoreline, Arvak, Great American, WQIS, Lloyds Gargrave EPG

---

[122] Alcantara, Leonard F./Cox, Mary A., *supra*, note 62, at 383, footnotes omitted.

[123] The OPA 90, §§ 2716(h).

[124] See 59 Fed. Reg. 34, 210 (1994), CERCLAR is the abbreviation of "The Comprehensive Environmental Response, Compensation and Liability Act of 1980".

[125] See 61 Fed. Reg. 9, 272 (1996).

[126] The information about "Shoreline Mutual" and "First Line programme of Stockton Re" is mainly from Bryant, Hugh, 'Specialist insurers offer real OPA solution for shipowners', *Lloyd's List*, Friday June 21 1996. Also see: Kim, Inho, 'Financial Responsibility Rules under the Oil Pollution Act of 1990', 42 *Nat. Resources J.* 565, at 577.

LLC, and Lloyds Gargrave UMS.[127] Moreover, 15 insurance companies have been approved to provide evidence of financial responsibility for all vessels.[128] Apparently, it was mainly due to the Clubs' refusal to be the guarantors under the OPA 90 that these new guarantors appeared. However, it has to be admitted that some of those special guarantors rely very much on P&I insurance. For instance, the Shoreline covers only liability which is not met by the P&I Clubs. Meanwhile, it has been observed that there remain complexities as regards the coordination between P&I Clubs and new financial responsibility guarantors with respect to oil-pollution insurance coverage and premiums.[129]

The US-style insurance market started to bring criticisms calling for legislation reform with regard to marine liability insurance. The view of a French insurance company appeared in the "Lloyd's List" in 2001.[130] It concerned tanker owners' liability insurance and pointed out that the cover provided by P&I Clubs fell far short of the emerging environmental risks. In addition, it called on the French government to press its European partners for a US-style approach to tanker liability, forbidding vessels from accessing European ports unless they could demonstrate that they had cover against environmental risks posed by cargoes they were carrying.

The current coverage offered by P&I Clubs is very beneficial to their members and at the same time facilitates the smooth operation of international maritime transportation. As shown above, external competitive pressure has emerged which the Club should not overlook. Apart from this, the Clubs should also be aware of their inherent shortcomings. At present, shipowners may have diverse insurable interests and trade in countries with extensively different liability regimes. It is possible that many shipowners can be exposed to liability risks that are not covered by P&I Clubs. Therefore, it may be necessary for P&I Clubs to introduce better means in order to offer insurance coverage for their members.

D. Concluding remarks

Shipowner may be required to maintain insurance or financial security to meet their potential liabilities on a national or international level. This is a main approach for covering shipowners' financial responsibility under both international civil liability conventions and the OPA 90.

The P&I Club is a mutual insurance association. "Mutual" means that the members contribute to a fund from which the losses or expenses suffered by them can be paid. Despite the emergence of other alternative means of providing

[127] Available at: <http://www.npfc.gov/cofr/graphs/image5.gif>, U.S. Coast Guard National Pollution Funds Center (visited 6 November 2005).
[128] Available at: <http://www.npfc.gov/cofr/graphs/image4.gif>, U.S. Coastal Guard National Pollution Funds Center (visited 6 November 2005)
[129] Kim, Inho, *supra*, note 126, p. 590.
[130] Spurrier-Paris, Andrew, 'Call for Reform of Liability Insurance', *Lloyd's List*, Monday, 8 January 2001.

insurance in the U.S. market, shipowners' oil-pollution liability is predominantly insured by P&I Clubs. The Clubs routinely provide insurance for liabilities not readily available elsewhere on the insurance market. Externally, the Clubs have established cooperative relationships with other organisations as regards handling oil-pollution incidents and related liability issues.

Chapter 3: Strict Liability and Insurance

A. Introduction

The shipowner is strictly liable for bunker-oil pollution in certain circumstances even in the absence of an international convention. If a bunker spill is from an oil tanker, the shipowner might be strictly liable for pollution damage sustained in a contracting State to the CLCs.[1] Some countries that base their national legislation on the CLCs have extended strict liability for oil pollution to non-tanker vessels.[2] More radically, some non-CLC States have established their own rules based on the strict liability principle regulating oil pollution from all types of ships.[3] However, except for a few countries that have adopted rules for bunker-oil pollution liability, claims for bunker-oil pollution have to be very much based on general liability rules on a national level. This may result in protracted and unsatisfactory litigation when serious oil pollution has occurred.

Inevitably, in consideration of the basis of liability for bunker-oil pollution, one tends to look to the law which has been developed both nationally and internationally to deal with similar risks arising from tanker oil spills. Strict liability for oil pollution damage was an innovative idea when it was introduced by the 1969 CLC. However, nowadays it has been accepted as the liability rule for various international conventions.

The nature of the liability regulated in Article 3(1) of the Bunkers Convention is essentially strict liability. Generally speaking, strict liability means that the shipowner is liable irrespective of the existence or lack of any fault on his part. Accordingly, in a bunker-oil spill incident, the shipowner will be liable only because his ship spilt bunker oil and caused pollution damage. However, it is not

---

1 See Chapter 1, Section B.I.: it is possible to hold the shipowner strictly liable for bunker oil pollution in limited circumstances under the CLCs.

2 The U.K. Merchant Shipping (Salvage and Pollution) Act 1994 provides "...by implementation of s.6 of the 1994 Act, the application of strict liability to non-tankers in ballast and non-seagoing vessels."

3 In the OPA 90, "vessel" is defined in §§2701(37) as meaning "every description of watercraft or other artificial contrivance used, or capable of being used, as a means of transportation on water, other than a public vessel." The enlargement of coverage of "vessel" in this provision accordingly extends the scope of the OPA 90 to cover bunker-oil spills. It is in order to provide certainty and improve the possibility of recovery of the cost of clean-up and preventive measures by victims from the bunker oil spill. See Chapter 2, Section E .II.2. a).

an absolute liability; the shipowner can be exonerated from liability when the situation falls within the exceptions in the Bunkers Convention.[4]

This chapter is mainly to discover the reasons for the establishment of the strict liability rule for bunker-oil pollution. An additional goal is to examine the relationship between the strict liability rule and compulsory insurance under the Bunkers Convention.

B. The basis of liability in the Bunkers Convention

Once the need for the Bunkers Convention was accepted, the question regarding the nature of liability was among the first to be considered. However, unlike the situation in the 1969 CLC, it did not involve so much debate on strict liability during the preparatory works of the Bunkers Convention due to the consideration that it was important to establish strict liability "for all forms of oil pollution". However, for the purpose of the topic of strict liability, it is worthwhile recalling the reasons why the strict liability rule was adopted in the CLCs and followed by the Bunkers Convention.

At the time of the *Torrey Canyon* incident, there was no internationally recognised liability rule specifically dealing with oil pollution. In addition, the national liability rule was inadequate or difficult to apply as a remedy for liability issues arising from this oil-spill incident. Since England was one of the major victims of the *Torrey Canyon* incident, the difficulties in handling issues regarding oil-pollution liability and compensation are often illustrated by detailing how English common law was utilised to deal with liability for pollution damage arising from this incident and other oil-spill incidents.

I. The difficulties in applying the common law of torts

At common law, when pollution damage was caused by an incident within U.K. territorial or internal waters, a claimant who suffered oil-pollution damage and consequential loss arising from such a spill might base his claim on: (a) trespass, (b) public or private nuisance, or (c) negligence. Each of these torts has particular requirements for recovery. Since this particular issue has been analysed by many scholars,[5] the following analysis, through referring to oil pollution in general, is also intended to demonstrate similar difficulties in applying those torts to bunker-oil pollution cases.

---

4 The Bunkers Convention, Art. 3(3), Art. 3(4).

5 For instance, Abecassis, David W.(ed.), *Oil Pollution from Ships* (1985), pp. 357-389; Gauci, Gotthard, *Oil Pollution at sea: Civil Liability and Compensation for Damage* (1997), pp. 10-16.

1. Trespass

"Trespass to land" is defined as "unjustifiable interference with the possession of land".[6] "Trespass is actionable *per se*, whether or not the plaintiff has suffered any damage."[7] In the incident such as an oil spill, the competent jurisdiction will take it as if the incident had occurred on land. In order to bring an action of trespass, the interference must be *direct* and the discharge *intentional*.[8]

However, due to the above-mentioned requirements, it is not easy to succeed in a claim for oil-pollution damage based on trespass. Firstly, it is difficult to prove that the incident was intended by the shipowner. Secondly, even if it is possible to prove that the oil was intentionally discharged at sea, it is not easy to prove that the spill or discharge of oil into the sea would lead to the contamination of the shore.[9] The argument from the shipowner will most likely be that the discharge of oil onto the foreshore was consequential or accidental, not direct or intentional. The weight of authority concerning tanker oil pollution in this respect is that it is unlikely that "trespass" will be allowed. It has been suggested that "…victims in a marine pollution case would not be able to bring an action in trespass since neither the direct consequence of the act, nor the intention to cause damage, could be established with ease."[10]

2. Nuisance

In *Southport Corporation*, Denning, L.J. held that: "if there is an unlawful interference with the plaintiffs' property, the question whether it is a trespass or a nuisance depends upon whether or not it is a direct physical interference."[11] In order to succeed in a claim on the basis of trespass, there must be unjustifiable interference with the plaintiff's property and the interference must be direct. However, if the interference is indirect, it may constitute a nuisance. In addition, to constitute a nuisance, the damage must be intentional. Nuisance is wider in scope than trespass. There are two kinds of nuisance: public nuisance and private nuisance.

---

6 Rogers, W.V.H., *Winfield and Jolowicz on Tort* (1998), p. 472.

7 See *ibid.*, p. 473, any footnotes omitted.

8 Wu, Chao, *Pollution from the Carriage of Oil by Sea: Liability and Compensation* (1996), p. 14.

9 Abecassis, David W.(ed.), *supra*, note 5, p. 359: "…there rarely will be sufficient certainty that the spillage or discharge of the oil on the sea will lead to the shore being contaminated-even in these days the wind and tide are not wholly predictable."

10 Wu, Chao, *supra*, note 8, p. 15, Abecassis, David W.(ed.), *supra,* note 5, pp. 358-363. Also see: *Southport Corporation v. Esso Petroleum Co.Ltd* [1956] A.C.218, H.L.; [1954] 2 Q.B.182, C.A.; [1953] 3 W.L.R.773, Q.B., *Fowler v. Lanning* [1959] 2 W.L.R. 249, Q.B., *Letang v. Couper* [1964] 2 All E.R.929.

11 *Esso Petroleum Co .Ltd Appellants; v. Southport Corporation Respondents.* [1956] A.C. 218 (H.L.), at 225.

Although there are some significant efforts made to define the limits of "nuisance", it is difficult to define it. It has been indicated that: "there is perhaps no more impenetrable jungle in the entire law than that which surrounds the word 'nuisance'...There is general agreement that it is incapable of any exact or comprehensive definition." [12] "Public nuisance" was explained in *Attorney General v. PYA Quarries Ltd.* in this way: "...It is ... clear, in my opinion, that any nuisance is 'public' which materially affects the reasonable comfort and convenience of life of a class of Her Majesty's subjects. The sphere of the nuisance may be described generally as 'the neighbourhood'..."[13] If an identifiable public right evidently exists, the interference with the public right will constitute a public nuisance. However, in the absence of a public right, an action for oil-pollution damage based on public nuisance may be relatively feasible only if: (1) there is a fact that a "neighbourhood" has been affected in a particular oil-spill incident; (2) a sufficiently large number of people have been affected by this incident; (3) the claimants can show that the damage suffered is particular and is different from the disturbances generally suffered by the public.[14]

A private nuisance, on the other hand, is an "unlawful interference with a person's use or enjoyment of land, or some right over or in connection with it."[15] In order to succeed in bringing a claim based on private nuisance, a claimant must have a proprietary interest in land and be able to prove that it has been interfered with.[16] However, most pollution victims have no property right in land at sea or along the coastal line. Therefore, it will be difficult for them to bring a claim on the basis of private nuisance.[17] Conversely, if victims can prove the existence of the necessary proprietary right, an action for private nuisance would be possible for the victims.

3. *Negligence*

Compared with the first two possible causes of action discussed above, negligence can be considered the most primitive. It is a widely used method devised to shift losses.[18] Negligence occurs when a person "has failed in his duty to take care of the interests of others [and] is liable for damage that is the foreseeable consequence of this dereliction of duty."[19] Therefore, for liability to arise in negligence, the following requirements have to be fulfilled: (a) there is a duty of care; (b) there is a breach of that duty of care; (c) the damage was caused by the breach; (d) the damage was not a remote consequence of the breach; and (e) there

---

12 Page Keeton, W. (gen.ed.), *Prosser and Keeton on the Law of Torts*, 5th ed., p. 616, 617.

13 *Attorney General v. PYA Quarries Ltd.* [1957] 2 Q.B. 169, at.170.

14 A detailed analysis is provided by Abecassis, David W. (ed.), *supra,* note 5, pp. 364-368.

15 Rogers, W.V.H., *supra*, note 6, p. 494.

16 Abecassis, David W. (ed.), *supra*, note 5, p. 369.

17 Wu, Chao, *supra*, note 8, p. 18.

18 Gauci, Gotthard, *supra*, note 5, p. 11.

19 Wu, Chao, *supra*, note 8, p. 17.

is foreseeability. It is also usual that only the former three elements are referred to as the main ingredients for negligence as a tort.[20] The burden of proof is on the plaintiff, who has to prove that the defendant did not take necessary or reasonable care to avoid the injury and that the damage caused was due to this breach of duty.

There is a general rule in law that "he who alleges must prove". The meaning of the rule is clear and the basis is evident. Nevertheless, the common law offers an exception, which is more popularly known by its Latin expression *res ipsa loquitur* (the thing speaks for itself). An explanation of this principle can be found in *Southport Corporation v. Esso Petroleum Co. LD. and Another.*,[21] where it states:

> "It is submitted that the principles applicable are correctly stated in Salmond on Torts, 11th ed., pp. 516, 517: 'The maxim res ipsa loquitur applies whenever it is so improbable that such an accident would have happened without the negligence of the defendant that a reasonable jury could find without further evidence that it was so caused '"

In other words, if the plaintiff puts forward reasonable evidence of negligence while the defendant cannot provide any reasonable explanation for the exception, the plaintiff will be entitled to succeed on the basis of negligence. This formulation obviously makes liability stricter as a practical matter. Thus, when this principle was applied in *Southport Corporation,* the House of Lords decided:

> "The Court of Appeal held that the doctrine of *res ipsa loquitur* applied, and that the onus was on the shipowner to explain why the steering gear went wrong and that, as they had not done so, they were liable in negligence."[22]

Therefore, the principle is applicable where it is shown that it is unlikely for the event to occur without negligence on the part of the defendant or some person for whom he is responsible. The application of *res ipsa loquitur* gives a considerable advantage to victims who have difficulty in finding evidence of fault.

In the case of oil pollution, however, it is questionable whether such a principle is useful for claimants. Certainly, its application depends on the facts of a particular case. Considering that well-navigated, well-maintained and properly operated ships are unlikely to collide with other ships or run aground, the principle may be applicable.[23]

From the foregoing discussion, one may conclude that it was not easy for oil pollution victims to succeed in a claim based on common law of torts. So far as English case law is concerned, the case *Southport Corporation*[24] is illustrative.

---

[20] Rogers, W.V.H., *supra*, note 6, p. 90: "Thus its (negligence) ingredients are: (1) a legal duty on the part of A towards B to exercise care in such conduct of A as falls within the scope of the duty; (2) breach of that duty; (3) consequential damage to B..."

[21] [1954] 2 Q.B.182.

[22] *Southport Corporation v. Esso Petroleum Co.Ltd* [1955] 2 Lloyd's Rep. 655.

[23] Abecassis, David W.(ed.), *supra*, note 5, p. 383.

[24] [1956]A.C.218, H.L.

The case is noteworthy because the *Inverpool*[25] went aground on December 3, 1950, but the House of Lords was able to deliver judgment only on December 12, 1955, which was after a long period of five years. This fact led Keeton[26] to note that, "even assuming that all the procedural problems could be solved, and the expenses of litigation met, an injured party who suffered a loss of his livelihood--or even a serious impairment of it--would have suffered extreme hardship in the interval and could have been ruined." Besides this, it was also doubtful whether victims could get full or adequate compensation for their damage even if they could successfully establish a claim. Therefore, it was deemed difficult under the ordinary rules of liability to ensure adequate protection and compensation owing to the potential scale of damage.[27]

II. Fault-based liability leads to unfair results for pollution victims

Fault-based liability is usually the liability for unacceptably unreasonable conduct. In order to judge that the defendant acted with fault, one needs first to determine the relevant standard of conduct and secondly to establish that the defendant failed to meet this standard.[28] The experience from dealing with tanker-related oil-pollution incidents shows that it is preferable to have strict liability rather than a fault-based liability for oil pollution.

One may argue that "the existing fault regimes, which include a mix of statutory and unwritten, customary rules and obligations, are at least as capable as strict liability regimes of dealing with complex situations…"[29] However, liability which is based on fault will leave too many gaps in the protection which should be afforded to the victims of oil pollution.[30] Liability based on fault means that the plaintiff must show fault on the part of the defendant or someone for whose acts the defendant is legally liable. In the first place, the concept of fault was quite likely to be interpreted in different ways by the courts of Contracting States and the uniform application of a civil liability convention could not therefore be expected.[31] Furthermore, in practice, maritime transportation is an activity over which potential victims have no control. Accordingly, it is often difficult to identify the tortfeasor and establish fault. As the case may show, it may be a very difficult task for non-marine-based or related victims to prove a marine fault. For

25 The incident involved a tanker bearing that name.

26 Keeton, G.W., 'The Lessons of the Torrey Canyon: English Law Aspects', *Current Legal Problems 1968*, Volume 21, pp. 94-112, at 98.

27 See IMCO: official records of the international legal conference on marine pollution damage, 1969, hereafter it will be called "O.R. 1969", at 627.

28 Werro, Franz/Palmer, Vernon Valentine (eds.), *The Boundary of Strict Liability in European Tort Law* (2004), p. 7.

29 Bergkamp, Lucas, *Liability and Environment: Private and Public Law Aspects of Civil Liability for Environmental Harm in an International Context* (2001), p. 264.

30 See IMCO, O.R.1969, *supra*, note 27, p. 460.

31 This was asserted by Professor Herber of the delegation of the Federal Republic of Germany. See O.R. 1969, *supra*, note 27, at 627.

instance, in order to find evidence, it is necessary for victims to acquire relevant information which is either inaccessible or the availability of which involves considerable costs. The information is, however, relatively easy for the shipowner to obtain.

Taking into account the facts that fault-based liability would result in unfair consequences, it was suggested that liability should be established on the basis of fault, with the burden of proof shifted to the shipowner in tanker-oil pollution incidents.[32] The reversed burden of proof is used in practice. It aims to reverse the burden of proof on to the owner by stipulating that the owner must prove that the pollution did not arise by any fault on his part. In other words, the owner will be liable for any pollution damage unless he can prove that the damage is caused "neither by any fault on the part of himself and his servants, nor by any fault in the operation, navigation or management of the ship committed by any person, whether or not his servant or agent."[33] This reversed burden of proof can relieve victims of the burden of proof and to some extent smooth the way for recovering compensation.

However, a reversal of the burden of proof in oil-pollution cases is insufficient to protect victims' interests. In many pollution cases, it is possible to establish fault on the part of the shipowner. However, in the complex context of shipping, oil-spill casualties may be caused by the fault of persons other than the owner of a ship. Therefore, where the owner claims that he was not at fault, extensive litigation will be involved, the cost of which must be borne by the victims. Due to the difficulty involved, it may consequently discourage victims from claiming compensation.[34] It is also possible that a ship may even founder in heavy weather and the oil may escape and cause pollution damage without any fault on the part of the vessel or his owner.[35] Therefore, imposing such a burden of proof on claimants may be unreasonable in certain circumstances.

Accordingly, liability based on fault with or without reversal of proof was not regarded as a sufficient guarantee for victims. The unfairness of fault-based liability may produce the following results: (1) the victims can get the compensation but only after the extensive and expensive claim procedure; (2) the victim can get the compensation but the compensation is inadequate; (3) the victim cannot get any compensation.

Due to the above-discussed difficulties, the consensus, right after the *Torrey Canyon* incident was that there was a need for a new liability regime, where those who suffered loss by oil pollution should be relieved of the extremely difficult task of establishing their claims and that the risk was to be borne by those industries that profited from maritime transportation. The strict liability rule was duly introduced in the 1969 CLC and remained in its 1992 Protocol. At the same

32 See IMCO, O.R.1969, *supra*, note 27, at 459.

33 See *ibid.*, at 458-459.

34 It may also be argued that it is unfair to require the shipowner to be liable if the pollution was not due to his fault. But under the Convention, there are exceptions to the liability such as in Article III (2) of the 1969 CLC.

35 See O.R. 1969, *supra*, note 27, at 461.

time, the adoption of the strict liability rule did not put the shipowner in an unprotected situation. He was entitled to be exonerated from liability in some circumstances and had a right of recourse independently of the Bunkers Convention.

III. Strict liability and its application

The landmark case in respect of strict liability is *Rylands v. Fletcher*, which was decided in England in 1968. Briefly, the fact was: the defendants, who were owners of a mill, had a reservoir constructed on nearby land. When the reservoir was being filled, the water leaked into connecting passageways and flowed into the nearby coalmines of the plaintiff. The defendants were not at fault because they had hired an independent contractor to build the reservoir. Therefore, the plaintiff brought an action of strict liability. The court thereafter held:

> "We think that the true rule of law is, that the person who for his own purposes brings on his lands and collects and keeps there anything likely to do mischief if it escapes, must keep it in at his peril, and, if he does not do so, is prima facie answerable for all the damage which is the natural consequence of its escape."[36]

Thus a stricter liability rule was applied in *Rylands v. Fletcher*. The principle enunciated in *Rylands v. Fletcher* was significant, since where a hazardous activity was involved, the party who created the risk and gained some economic benefit from that activity should bear the risk.[37] It has, nevertheless, been indicated that there is no modern example of open judge-made strict liability in the English common law, at least in the sense of the creation of a category of liability.[38] However, "with passage of time and advancing knowledge of risk and perception of society's demands for protection from them",[39] this type of strict liability rule has been extended to be used in many aspects in English Law, such as in the legislation regarding oil pollution at sea. It is a standard of liability in a much stricter sense and it can offer better protection to innocent victims.

No matter what it is called or whatever form it takes, many countries have gradually introduced a strict liability rule in their national legislation.[40] For

36 [1868] Lloyds' Rep. 3 H.L. 330. Also see Vandall, Frank J., *Strict Liability: Legal and Economic Analysis* (1989), pp. 1-12.

37 Wilde, Mark, *Civil Liability for Environmental Damage* (2002), p. 43, 44.

38 See "England" part by Rogers, W. V. Horton, in: Koch, B.A./Koziol, H. (eds.), *Unification of Tort Law: Strict Liability* (2002), p. 108.

39 See *ibid.*

40 Koch, B.A./Koziol, H. (eds.), *ibid.*, pp. 395-396: "As a first observation, it is necessary to point out that strict liability in most jurisdictions predominantly seems to be based on singular rules rather than general or at least broader clauses. This is particularly noteworthy for civil law countries. While Austrian courts, for example, (cautiously) apply existing strict liability laws analogously (which reduces problems of tardy), German and Swiss practice so far deny the possibility of extending such statutory rules at all. French law, on the other hand, not only has a

instance, in Germany, separate statutes introduced strict liability, *inter alia*, for motor vehicles (in 1908), aircraft (in 1936), nuclear installations (in 1959), facilities causing pollution of water (in 1957) and manufacturers of pharmaceutical drugs (in 1976). The German courts have refused to extend strict liability by analogy beyond the specific cases defined by the legislature.[41] In fact, the risk specific approach in Germany has influenced many other countries with the civil law system. In France, there are many strict liability legislations both in public and private law, some of which apply to damage caused by things and others to damage caused by specific activities.[42] For example, as far as private law is concerned, the courts have developed a rule according to which the custodian of a thing is, irrespective of fault, liable for damage caused by the thing.[43] Several French statutes attach strict liability to such activities as transportation of oil by sea, transport of passengers – whether by road, air or sea – and so on. Strict liability is a liability rule based on risk, not on fault.

The above examination, although not exhaustive, shows the extended application scope of the strict liability rule. This writer notes, however, that a number of texts use the term "absolute liability" instead of "strict liability". "Strict liability" and "absolute liability" should not be understood as synonymous concepts. Strict liability allows the taking into account of factors such as the conduct of the third party and treats some circumstances as exceptions. In contrast, absolute liability is a type of liability for the damage arising from an activity where no account is taken of the standard of care exercised by the responsible party and no defences to liability are provided.[44] Absolute liability is imposed in respect of risks that can cause catastrophic harm, such as the exploitation of nuclear energy.[45]

---

clause which (at least in today's understanding) introduces general liability for 'deeds of the things within one's keeping' (the famous Art.1384 subs. 1 Code Civil), courts furthermore seem to be quite open for an extensive application of other rules (such as the loi Badinter)." Citations omitted.

41 Pfennigstorf, Werner/Gifford, Donald G., *A Comparative Study of Liability Law and Compensation Schemes in Ten Countries and the United States* (1991), p. 51.

42 See 'France', by Galand-Carval, Suzanne in: Koch, B.A./Koziol, H. (eds.), *supra*, note 38 , pp. 127-145.

43 Bocken, H., 'Developments with respect to Compensation for Damage Caused by Pollution', in: Markesinis, B.S. (ed.), *The Gradual Convergence: Foreign Ideas, Foreign Influences, and English Law, on the Eve of the 21st Century* (1994), pp. 226-251, at 233.

44 See IMO LEG 74/4/3, the proposal submitted by Greenpeace International to the Bunkers Convention.

45 To found a claim for nuclear damage, it is necessary for a claimant to prove that such damage has been caused by a nuclear incident involving the ship. The operator of the ship is then liable irrespective of whether his or a third party's fault or negligence has caused the nuclear incident.

C. The main reasons for introducing strict liability

I. Ensuring protection of and compensation for victims

The possible seriousness of pollution damage in an oil-spill incident justifies the establishment of a separate liability regime and also forms the basis for the imposition of strict liability for oil spillages. Therefore, if a causal link can be established between the oil-spill incident and pollution damage, the shipowner shall be liable for pollution damage, regardless of whether the pollution was caused by the negligence or other known or unknown reasons. The strict liability rule was chosen since it was under a situation "where it has been determined that the need to…provide adequate compensation outweighs the need to establish any form of moral culpability on the part of the defendant."[46] Imposing strict liability on the shipowner means that it is important for victims to obtain compensation once pollution damage has occurred. Strict liability "…like the rule of constructive total loss in marine insurance, is itself a legal fiction which is probably justified because the benefits outweigh the possibility that the resulting loss-allocation is not compatible with principles of justice."[47] Over the years, it has shown its efficiency and allows different categories of victims to claim for compensation. This might also be the basic reason for adopting the strict liability rule in the Bunkers Convention.

However, as far as the desired purpose of ensuring compensation or protection is concerned, there are limits under the Bunkers Convention: first, the types of pollution damage are specifically defined.[48] Any victim will not be compensated unless he suffers any pollution damage as defined in the Bunkers Convention. In addition, pollution damage, for instance, caused by any natural phenomenon of an exceptional, inevitable and irresistible character cannot be compensated.[49]

Secondly, if strict liability is to ensure compensation for pollution victims, it is relevant to ask: against whom may the victims file a claim if the shipowner is insolvent after an incident in the absence of the established financial security? A strict liability rule in such an unfortunate case would be pointless. The requirement of compulsory insurance to some extent relieves the hardness of such a situation. But, in addition, as one author notes, "strict liability by itself also suffers from a very serious limitation which afflicts any system of compensation which is based on the law of torts: the system is effective only as long as the defendant or his insurer or guarantor is willing or can be forced to pay."[50] In other words, strict liability imposed on the shipowner is useless unless he is capable of paying compensation or taking out liability insurance to cover his liability and the provider agrees to pay compensation if the liability can be established.

---

46 Wilde, Mark, *supra*, note 37, p. 197.

47 Gauci, Gotthard, *supra*, note 5, p. 19.

48 The Bunkers Convention, Art.1(9).

49 The 1969 CLC, Art. III(2). For more detailed discussion about pollution damage, see Chapter 6, Section E.

50 Gauci, Gotthard, *supra*, note 5, p. 20.

Thirdly, the alleged compensating benefit is in question in that liability is limited under the Bunkers Convention. By contrast, under the general fault regime of tort law, no limitations apply and the victim is entitled to full compensation.[51] Therefore, the limitation amount which has to satisfy the claims of compensation is crucial.

II. The impact on the industry

1. The industry bears the cost of pollution damage

The industry here refers to the shipping and cargo industry. Due to the complexity of the variables involved in maritime transportation, the industry is in a better position to bear pollution damage for the following reasons.

Firstly, the industry is in a better position to access technical and technological information. The shipowner is regarded as the party who has superior knowledge of the navigation and management of transportation. The cargo owner, when chartering a ship, is presumed to obtain as much information as possible if only to protect his company's reputation. In comparison, the victims usually have little or no idea about the potential damage and have no knowledge of how to protect their interests. In short, the operation of the industry is within its own control and, consequently the means and methods of preventing pollution damage are also within its influence. It therefore makes sense to hold the relevant industry liable rather than allow innocent victims to suffer.

Secondly, the industry is in a better position to absorb the loss arising from pollution damage. On the one hand, it is logical that the industry that benefits from sea transportation shall also accept the risk it entails. On the other hand, although shipowners are liable for pollution damage, they can partly shift the burden arising from the said damage to other parties. For instance, they may raise the price of the charter party in order to transfer one part of costs of the potential damage, and the costs of damage will thus be distributed among relevant parties, such as the charterers, cargo-owners, or even society as a whole.[52]

In addition, the fact that the industry is in a better position to bear pollution damage is also due to the availability of liability insurance for them to cover the risks. The availability of insurance plays an important role in deciding which party should bear the liability. In reality, it may be possible for victims to insure themselves. The difficulties, however, consist in the complexity and difficulty of

---

51 Faure, Michael (ed.), *Deterrence, Insurability, and Compensation in Environmental Liability: Future Developments in the European Union* (2003), p. 30.

52 Strict liability imposed on the industry is particularly advantageous in the context of an international and uniform regulation. Otherwise, for instance, the shipowner, who may be subject to strict liability at the local level will be confronted with a competitively disadvantageous position and will even be in danger of going out of business. The adoption of this liability rule at international level will, however, subject all shipowners to it. No shipowner will therefore have any disadvantage over others.

defining "pollution victim" in maritime transportation. In the circumstance of an oil-spill incident, pollution victims might be an incoherent group and of different types. In addition, due to the fact that the types of pollution damage suffered will be a matter of fact and different victims may suffer different types of pollution damage, the types of pollution risk against which the victims should insure themselves are not certain either. Many victims might even have no sense of possibly suffering pollution damage and therefore of the need to purchase insurance. In fact, it is also not clear whether the kind of insurance is available to victims. By contrast, the shipowner usually purchases liability insurance to cover his potential third-party liabilities. It is, therefore, more convenient for him to be liable and purchase insurance to cover oil pollution damage.

2. *Incentive to improve prevention of marine pollution*

"Deterrence" is viewed as an important objective of liability. The Dutch scholar, Wertheim, who used the "activity" theory, identified this "general deterrence" objective in 1930.[53] "Deterrence" means that a liability regime should be able to provide incentives for the operators to use due care and reduce the risks associated with that activity. In other words, a liability regime must deter from exposing other persons to excessive and unreasonable risks.[54]

Strict liability may impel the shipowners to prevent marine pollution in that it may create an incentive on the part of shipowners to comply with the standards and take the required measures in order to avoid liability.[55] Since the liability rule shifts the damage from victims or persons initially suffering from it to the shipowner, whether he is at fault or not, the shipowner will thus exercise a certain degree of care to avoid the incidents of liability. In addition, should he not be able to bear the loss and go out of business, it will be an example to other shipowners, who will learn to exercise more care. Shipowners may, *inter alia*, develop stringent methods to ascertain the seaworthiness of ships, undertake preventive measures, hire more qualified seamen and also develop some other precautionary measures. In this way, strict liability rule may provide an incentive for the shipowner to use due care and reduce potential risks associated with maritime transportation, or even remove substandard ships from international service.

It is important to note that the primary responsibility for the enforcement of international rules and standards, which have established to prevent, reduce and control marine pollution from vessels, lies generally with the flag State[56] and not with the shipowner and so the latter has no direct incentive to prevent pollution.

---

[53] Wertheim WF. Aansprakelijkheid voor schade buiten overeenkomst. Dissertatie. Leiden. 1930. Quoted by Bergkamp, Lucas, *supra*, note 29, p. 87.

[54] For more details, see Bergkamp, Lucas, *ibid.*, pp. 86-96. For more discussion regarding the issue that strict liability is preferred, see Section B.II in this chapter.

[55] Wolfrum, Ruediger/Roeben, Volker/Morrison, Fred L., 'Preservation of the Marine Environment', in: Morrison, Fred L./Wolfrum, Rüdiger (eds.), *International, Regional and National Environmental Law* (2000), pp. 225-283, at 268.

[56] Wolfrum, Rüdiger/Roeben, Volker/Morrison, Fred L., 'Preservation of the Marine Environment', in: Morrison, Fred L./Wolfrum, Rüdiger (eds.), *ibid*, at 264.

By contrast, strict liability under international civil liability conventions is imposed on the shipowner; in addition, the relatively high limit of liability, once any liability is incurred to the shipowner, puts the pressure on the shipowner to avoid his liability to the minimum.

One may argue that fault-based liability may provide a superior incentive for complying with regulatory standards than strict liability.[57] Shipowners might be less concerned about the prevention of pollution under a strict liability rule since they have to be liable except in exonerating circumstances. In effect, both liability rules provide the shipowner with certain incentives to take efficient care, since taking efficient care will minimise the expected incident costs which the shipowner has to bear under any liability rule.[58]

To sum up, it might be controversial whether the strict liability rule can provide a better incentive for pollution prevention or incident reduction. However, it is certain that strict liability can facilitate victims' claims.

D. Distribution of liability and exceptions to liability

To establish a standard of liability for oil pollution, one should also take into account the interests of the relevant liable person. This section will focus on the issue of the distribution of liability and the exceptions available to the liable person.

I. Who shall be liable?

Any liability rule channels the obligation to bear a loss to an identified category of persons.[59] It is already clear that the shipowner must be strictly liable for oil pollution under the Bunkers Convention. However, one is prompted to ask why it is the shipowner and not the cargo-owner or other parties involved in maritime transportation who must be held liable. The dispute on this issue was much fiercer in the case of tanker oil pollution because the risk was considered to derive both from the particular form of carriage and from the nature of the cargo. All the views expressed on this matter can be generally classified into two groups imposing: (1) liability on cargo interests, and (2) liability on shipping interests.

---

57 Landes, W.M. and Posner, R. A., *The Economic Structure of Tort Law* (1987), p. 259.

58 Faure, Michael, 'Economic Analysis', in: Koch, B.A./Kozio, H.(eds.), *supra*, note 38, p. 364, 365.

59 Bergkamp, Lucas, *supra*, note 29, p. 70, FN.21.

1. Provisions in the CLCs

This discussion will primarily refer to the provisions in the CLCs and the proceedings in its enactment. The analysis is based on the records of the discussions about the 1969 CLC.[60]

a) Liability of the cargo interest

The proposal to impose liability on the cargo was mainly due to the nature of the cargo transported. It was the cargo that caused the damage and not the ship.[61] The cargo-owner should necessarily insure pollution liability or undertake pollution liability in another form. In addition, such a liability imposed on the cargo-owner would act as a motivation for him to select the best ship and choose the safest route for his shipment. Furthermore, the cargo-owner was considered to be the party who stood to profit most from maritime transportation and was financially able to pay compensation.

However, there were some practical difficulties. The most significant one was that the identity of the cargo-owner could be changed or shifted during a voyage, since the ownership of the cargo is often subject to the bill of lading and varies depending on what the bill of lading provides.[62] It was thus difficult for victims to identify the liable person. This fact was also incompatible with the envisioned "compulsory insurance" provision, since the flexibility of changing cargo ownership would increase the difficulty in determining how to carry out an insurance scheme and how to ensure the validity of the insurance certificate.[63]

If the liability was to be imposed on the cargo, it should be imposed on the cargo interest most easily identifiable. The cargo interests might include a variety of persons such as the shipper, the receiver or the owner for the time being of the cargo. Among the representatives in the discussion, the Irish delegation was the strongest supporter of the proposal to impose liability on the shipper. Generally, in shipping law, the meaning of "shipper" depends upon the context in which it arises. The courts have used the term to refer to the party who contracts with the carrier.[64] The shipper was considered to be "a constant factor known to the shipowner even if not to the victim".[65] In the Irish proposal, the burden of identifying the shipper was placed on the shipowner. If the shipowner failed to identify the shipper, the shipowner must stand in the shoes of the shipper whom he had failed to identify. In this event the shipowner would pay compensation for pollution damage not so much as the shipowner but as the constructive shipper.[66] In this system, the shipowner's liability was based on fault, while the shipper was required to be strictly liable. If the ship were at fault, the cargo-owner would

60 See IMCO, O. R.1969, *supra*, note 27.
61 See *ibid.*, at 635.
62 See *ibid.*, at 458.
63 See *ibid.*, at 458.
64 Baughen, Simon, *Shipping Law* (2004), p. 8.
65 See IMCO, O.R., 1969, *supra*, note 27, at 635.
66 See *ibid.*, at 537.

recover part of what he had paid from the ship; if there were no fault, the cargo-owner would not recover anything.

The whole proposal was constructive. It was based on the considerations of equity between the ship and the cargo. However, as Lord Devlin[67] pointed out:

> "...the best channel for providing compensation was one which was simple in application, made use of existing procedures and offered an incentive to prevent casualties. The Irish proposal did not provide the simplest channel. The easily identifiable shipowner was replaced by the shipper, whom the shipowner would be required to identify. More than one person had to insure (the shipper for his liability under the Convention and the shipowner both for his normal liability and also against the default by the shipper)..."[68]

In addition, there was also the consideration that the shipper was not always the real cargo-owner.[69] And even the fact that there might be more than one shipper would considerably complicate the limitation of the shipper's liability.[70] Therefore, in practice, it would be difficult to impose liability on the shipper.

b) Liability of the ship

The main reason of imposing liability on the shipowner is that the shipowner or the operator is usually in a position to preclude or reduce to a minimum the risks arising from the carriage of goods. Moreover, it is easier to identify the shipowner than the representatives of the cargo interests. Different opinions were presented during the discussions of the new liability regime which focused on whether to place the liability on the operator or on the shipowner.

An operator was defined as "a person using in his own name the ship manned, equipped and supplied by him."[71] He was the person who can either cause or avoid damage by the management of his ship.[72] However, it was pointed out that the term "operator" did not actually belong to the field of maritime law because it had been borrowed from the nuclear conventions. In the nuclear industry, a person who had received authorisation from a State to operate was the "operator" and this person was easily identifiable. In contrast, the "operator" in the shipping industry would be more difficult to determine.[73] Furthermore, it was considered that "any liability on the operator would be incompatible with the envisaged scheme of compulsory insurance, since countries would have to continually issue and revoke insurance certificates as the terms of a charter changed."[74] Under these considera-

---

67 Soon after the *Torrey Canyon* incident, the CMI established an International *Torrey Canyon* Sub-Committee and Working Group, which was chaired by Lord Devlin, to work in co-operation with IMCO on the private law aspects.

68 See IMCO, O.R. 1969, *supra*, note 27, at 638.

69 See *ibid.*, at 640.

70 See *ibid.*, at 641.

71 See *ibid.*, at 443.

72 See *ibid.*, at 690.

73 See *ibid.*, at 444.

74 See *ibid.*, at 690.

tions, one may inevitably conclude that "a liability regime imposed on the operator will be more of an obstacle than a solution."[75]

It was of primary importance that there must be a clear identification of the liable person to ensure with absolute certainty that the victims would be properly compensated. The choice of the shipowner seemed to be the most feasible one. Eventually, the delegations decided to impose liability on the shipowner. The "shipowner" is defined as "the person or persons in whose name the vessel is registered or, in the absence of registration, the person or persons who own the vessel..."[76] Imposing liability on the registered owner would make it easier for claimants to discover the liable party.

It was for the reasons cited above that the 1969 CLC channelled liability not to the cargo-owner, nor to the operator of the vessel, but to the shipowner.

c) The second-tier liability of the cargo-owner

After the protracted discussion preceding the 1969 CLC, it was agreed upon, as a compromise, that the shipowner was to be strictly liable. This was brought about by an agreement that the shipowner's liability for oil pollution would be complemented by an international compensation fund financed by oil companies. Accordingly, the interests involved were balanced through the shift of a substantial portion of the compensation burden onto the oil industry, one of the main beneficiaries of the carriage of oil by sea.

In practice, the cargo interest creates a second-tier fund, from which the compensation will be paid when the total aggregate of claims resulting from one incident exceeds the shipowner's liability limit or when, for whatever reason, compensation cannot be obtained from the shipowner.[77] The whole system is regulated under the Fund Convention. More precisely, instead of imposing liability on individual cargo-owners, the Fund Convention provides for that compensation to be paid by an intergovernmental body established to administer the Fund, and this Fund is financed by contributions levied on the oil importers in the contracting States.[78] The obligation to finance is not imposed upon the exporter of oil, since it was acknowledged that any export is also an import to someone else.[79] Nor is the contribution to the Fund imposed on all oil imports due to the consideration that it would be uneconomic to include small importers for

---

75 Wu, Chao, *supra*, note 8, p. 54.

76 The 1969 CLC, Art. I(3).

77 For instance, the Fund is liable when the incident is caused by an Act of War, intentional act of a third party or negligence of the government. The shipowner is exonerated from liability in these three situations.

78 Art. 10(1) of the Fund Convention provides that: "Contributions to the Fund shall be made in respect of each Contracting State by any person who, in the calendar year referred to in Article 11, paragraph 1, as regards initial contributions and in Article 12, paragraph 2(a) or (b), as regards annual contributions, has received in total quantities exceeding 150,000 tons:..." For more, see Chapter 1, Section A.II.

79 Wu, Chao, *supra*, note 8, p. 95.

whom the administration cost for the IOPC Fund will be greater than the amount of their contributions.[80]

The oil industry, as the cargo interest, thus plays a great role in compensation through the development of a supplementary compensation source. The oil industry's second-tier compensation contributes to the shipowner's willingness to be strictly liable for pollution damage.

2. *Liability rule under the Bunkers Convention*

Liability of the shipowner is stipulated in Article 3 of the Bunkers Convention. There are two groups of persons who might be liable: (i) liable persons under Article 3(1) referring to "shipowner", which includes the registered owner, bareboat charterer, manager and operator of the ship.[81] The persons in the definition of "shipowner" as a whole are liable. This was based on the idea that increasing the number of liable persons would enhance the availability of compensation for oil-pollution victims; (ii) the liability insurer or guarantor meaning a bank or similar financial institutions which will cover the liability of the registered owner for pollution damage.[82] The liability insurer, in the light of the direct action provision of the Convention, is liable for the registered owner's liability for pollution damage.[83]

Apparently, liability is not imposed on the cargo-owner. The owner of the cargo is not strictly liable for pollution damage under the Bunkers Convention. The cargo-owner is strictly liable in a tanker oil spill incident, since pollution damage originates from the nature of the cargo. However, the situation is different when relating to the carriage of other types of goods by sea. It may be a type of regular cargo, not even hazardous in nature, so it is not reasonable to impose strict liability on the cargo-owner under the Bunkers Convention. Nevertheless, although the cargo-owner is not required to be liable, the shipowner, depending on his bargaining power, may make the cargo owner liable for pollution damage under the charter party or under the contract of carriage of goods by sea.

II. Exceptions to liability

The exceptions to liability usually exist in any liability system in order to relieve the liable person from liability.[84] The strictness of the liability rule can be raised or lowered depending on the list of exceptions which may be available for the liable person. Exceptions exist in the cases of strict liability for oil-pollution damage.

---

[80] *Ibid.*

[81] The Bunkers Convention, Art. 1(3).

[82] The Bunkers Convention, Art. 7(1).

[83] The Bunkers Convention, Art. 7(10).

[84] Morrison, Fred L./Wolfrum, Rüdiger (eds.), *supra*, note 55, p. 832: "Once a system has established a standard of liability and an actor within that system fails to meet that standard, exceptions to liability may still be possible..."

Once the liable party proves that the damage resulted from an exception, he will not be liable for pollution damage.

If the liability is voluntarily or compulsorily insured, the insurer becomes liable under the insurance policy. The exceptions will also be applicable to the insurer. The insurer is entitled to enjoy the exceptions to liability which the shipowner would have been entitled to enjoy. Furthermore, where the insurer is sued directly, he may additionally invoke such defences as wilful misconduct of the shipowner, which the latter cannot of course invoke. The following sections will focus on the illustration of the exceptions which are applicable to both the shipowner and his liability insurer; the defences available only to the insurer when sued directly will be discussed in Chapter 6.

1. Types of exceptions in general

Generally speaking, strict liability, but not an absolute liability,[85] explicitly or by reference to the general law, provides for exceptions to liability. They are defences for the shipowner and his insurer to avoid liability. The provisions in respect of the exceptions from liability may vary widely in different liability regimes. In general, the liable person can be exonerated from his liability under the following circumstances in any legal system: (1) act of God, *force majeure*, unavoidable event; (2) fault of victims; (3) fault of a third party; (4) other defences.[86]

An "act of God" is of limited practical importance. In English law, the meaning is limited and "...It does not mean an exceptional natural event nor even one which could not reasonably have been anticipated at that time, but one of which human prudence would not even recognise the possibility – for example, a serious earthquake or tidal wave in England. But not the great storm of 1987, nor the great French storm of 1999..."[87]"*Force majeure*", under French Law, is defined as "a natural event or act of either the victim or a third party, which was external to the defendant (i.e. outside its sphere of activity), unforeseeable and unpreventable,"[88] the conditions of which are cumulative. In Germany, strict liability is excluded if *force majeure* can be proved (so called '*hoehere Gewalt*'). As in the case of an unavoidable event, *force majeure* must represent an external influence which cannot be avoided even with utmost care.[89] It is therefore safe to say that in different countries, there might be differences in the application of these concepts.

As far as "fault of victims" is concerned, it is usually not considered as a complete defence; it depends on the facts of the case. If the damage is wholly caused by the fault of the victims, the liable person can be exonerated from liability completely. However, most systems of legislation contain a contributory negli-

---

85 See Section B.II of this chapter, absolute liability is that which does not provide any grounds for exoneration or any defences.

86 These categories are based on the questionnaires in: Koch, B.A./Koziol, H. (eds.), *supra*, note 38.

87 Rogers, W. V. Horton, 'England', in: *ibid.*, pp. 101-126, at 118.

88 Galand-Carval, Suzanne, 'France', in: *ibid.*, pp. 127-145, at 137.

89 Fedtke, Joerg/Magnus, Ulrich, 'Germany', in: *ibid.*, pp. 147-176, at 163.

gence rule, which, when applied, will only result in a corresponding reduction in the amount of damages awarded. The notion of contributory negligence means that any relevant participation of victims in the damage or harm is relevant and should be taken into account when assessing damages. If contributory negligence is found, the damages for victims must be reduced proportionately or may even be excluded. The defendants, however, are obliged to prove the fault on the part of the victims or their participation in causing the damage or harm.

Legislation from different municipal jurisdictions reveals different views suggesting "fault of third party" as an exception against liability. For example, in Germany, the fault of a third person generally cannot affect strict liability.[90] But in the Netherlands, the fault of a third party will generally lead to a joint liability.[91] In other countries, there is such a rule as follows: if a third person's tortious conduct was a decisive cause of the loss, the liability is to be denied; or if a third party's fault can be connected with an actual portion of the damage, then the latter has to be split accordingly.

The category of "other defences" provides other specific grounds for exonerating a shipowner from liability. These grounds are expressly provided in different statutes.

2. *Exceptions available to the shipowner in the Bunkers Convention*

The exceptions are included in Articles 3(3) and 3(4) in the Convention. Article 3(3) reads:

> "No liability for pollution damage shall attach to the shipowner if the shipowner proves that:
> (a) the damage resulted from an act of war, hostilities, civil war, insurrection or a natural phenomenon of an exceptional, inevitable and irresistible character; or
> (b) the damage was wholly caused by an act or omission done with the intent to cause damage by a third party; or
> (c) the damage was wholly caused by the negligence or other wrongful act of any government or other authority responsible for the maintenance of lights or other navigational aids in the exercise of that function."

The above sub-paragraph (a) includes risks which rarely occur in practice. However, once they occur, they will cause losses of such a magnitude that liabilities of the insurer are, in consequence, likely to be increased far beyond what they would be in the ordinary course of business. The liability insurer is usually not willing to insure such liabilities. For instance, P&I insurance generally excludes from its coverage costs and expenses arising out of or consequent upon a war, civil war, rebellion, revolution, civil strife, insurrection or any hostile acts. However, there are policies which can be drafted to cover specific war risks.

The sub-paragraph (a) further excludes the liability resulting from "a natural phenomenon of an exceptional, inevitable and irresistible character". This is more

---

90 *Ibid.*, at 164.

91 Perron, Edgar du/Boom, Willem H. van, 'Netherlands' section, in: *ibid.*, pp. 227-255, at 147.

generally known as an "Act of God", *force majeure* or unavoidable event, which we have discussed in the above section. In order to relieve the shipowner of his liability, the natural phenomenon must be of an exceptional, inevitable and irresistible character according to the Bunkers Convention. "Inevitable" is a key factor, since "it is a cardinal principle that insurance indemnifies against accidents rather than certainties."[92] It has been suggested that, for the exception to apply, the defendant must show that "in no circumstances could anyone have avoided the accident."[93] It is not sufficient if he merely shows that he did his reasonable best.[94] It also appears that the condition of being "exceptional, inevitable and irresistible" must exist at the same time. Moreover, "it seems clear that the phrase does not cover hurricanes, for these are negotiable by some ships, but it would cover tidal waves."[95]

The sub-paragraph (b) refers to the act or omission of a third party. It is important to clarify three points relevant to this sub-paragraph: first, a shipowner can be exempted from liability if pollution damage is wholly caused by something done or left undone by the third party. Secondly, the concept of "third party" involves persons not being "the registered owner, bareboat charterer, manager and operator of the ship"[96] and their servants or agents. It may also exclude pollution victims in this context due to an independent paragraph in Article 3(4).[97] Thirdly, the third party's conduct is the sole cause of the spill or discharge of oil. However, if a shipowner cannot show that the third party's conduct was with the intent to cause pollution damage, he cannot exempt himself from liability. Accordingly, it will operate to relieve the shipowner of his liability only if any damage was wholly caused by an act or omission done with the intent to cause damage by a third party. For example, "barratry" was considered to fall within the exception, because barratry is defined in UK insurance law as "every wrongful act wilfully committed by the master or crew to the prejudice of the shipowner". A member of the crew becomes a third party in committing a "frolic of his own"[98] through the performance of an act of barratry.[99] It is in line with the requirements in the exception provision. The liability of the third party will not be governed by the Bunkers Convention. In other words, if the fault of the third party was the sole cause of an incident, the victims can claim against the third party only under ordinary tort law.

The sub-paragraph (c) gives the shipowner a defence if he can prove that a negligent or wrongful act of a relevant government or other authority caused the oil spill. This authority must be the agency that is responsible for the maintenance

---

92 Bennett, Howard, *The law of marine insurance* (1996), p. 229.

93 Forster, Malcolm, 'Civil Liability of Shipowners for Oil Pollution', *J.B.L.* (1973), pp. 23-31, at 26.

94 *Ibid.*, pp. 25-26: "If another could have succeeded in averting the occurrence, then the occurrence is not inevitable and there is no defence."

95 Abecassis, David W.(ed.), *supra*, note 5, p. 205.

96 The Bunkers Convention, Art.1(3).

97 The Bunkers Convention, Art. 3(4).

98 *Joel v. Morison* (1934) 6 C&P 501.

99 More detailed analysis see Gauci, Gotthard, *supra*, note 5, pp. 75-76.

of lights or other navigational aids. The shipowner is exempted from liability even if the act was conducted unintentionally.

Besides the above-mentioned provision, Article 3(4) also provides for contributory liability exemption. As a sub-paragraph, it provides:

> "If the shipowner proves that the pollution damage resulted wholly or partially either from an act or omission done with intent to cause damage by the person who suffered the damage or from the negligence of that person, the shipowner may be exonerated wholly or partially from liability to such person."

It explicitly mandates that if the shipowner proves that the pollution damage resulted wholly or partially either from an act or omission by victims with intent to cause damage or from the victim's negligence, the shipowner may be exonerated wholly or partially from liability to such person. In other words, according to the facts of the case, all the factors contributing to the incident resulting in pollution damage to victims shall be weighed up and the relevant proportion of the liability will thereafter be assessed. Under this system, the shipowner may assume liability wholly or only partially. This provision is narrower and more specific than Article 3(3)(b). This provision, like the similar one in the CLCs, "was introduced on grounds of fairness, despite anxieties on the part of some that it would weaken the concept of strict liability."[100]

It is important to note some phrases in the above-listed paragraphs. These include the phrases of "resulted from", "wholly caused" and "resulted wholly or partially from…" They have different meanings, as they can either mitigate the shipowner's liability or completely exonerate him from liability.

The phrase "resulted from" is used in Article 3(3)(a). The meaning of this phrase is quite controversial. Examinations of the phrase in the similar provision of the CLCs show that there are two main opinions. One of those points out that it is not necessary for pollution damage to be "wholly caused" by, for instance, "an act of war, hostilities, civil war, insurrection or a natural phenomenon of an exceptional, inevitable and irresistible character".[101] It is sufficient that any act or natural phenomenon falling within the provision is the proximate or dominant cause of pollution damage and other contributory factors need not be considered. For example, crew negligence or other factors will have no bearing on the state of liability despite being contributory causes.[102] This opinion emphasises the "proximate or dominant" nature of the relevant facts of the act of war, etc. Meanwhile, one author has another opinion. She opines that: "The term 'resulted' is vague…It can therefore be reasonably concluded that the presence of an act of war or *force majeure* alone is sufficient to exonerate the shipowner from his liability, however important the role played by the *force majeure* or act of war."[103] Obviously, this opinion only emphasises the existence of the relevant facts of the

[100] Abecassis, David W.(ed.), *supra*, note 5, p. 206, the author quoted IMO documents: LEG/CONF/C.2/WP.41,OR 601, LEG/CONF/C.2/SR.18,OR 738.

[101] The 1969 CLC, Art. III (2)(a).

[102] DeLaRue, Collin M./Anderson, Charles B., *Shipping and the Environment* (1998), p. 88.

[103] Wu, Chao, *supra*, note 8, p. 61.

"act of war, etc.", which may not be the exclusive or dominant reason for bunker-oil pollution. Thus, it is questionable whether there is a need to identify the attendant circumstance as being the "proximate and dominant cause" of the pollution incident. Unfortunately, the Bunkers Convention is not clear about this point.

However, the uncertainty may be clarified if we examine the phrase "resulted from" in the context of marine insurance. Under the principles of causation[104] in English marine insurance law, the phrase 'arising from' has been held to be synonymous with the "proximate cause" test. The "proximate cause" of the loss[105] has been discussed in many House of Lords decisions as not mainly referring to proximity with regard to time. Lord Shaw, of the House of Lords, stated in *Leyland Shipping Co. v. Norwich Union Fire Insurance Society*[106] that: "...To treat proxima causa as the cause which is nearest in time is out of the question".[107] Furthermore, Lord Wright, in another case,[108] gave the opinion that "'proximate' here means, not latest in time, but predominant in efficiency." In the same case, Lord Wright further held: "...This choice of the real or efficient cause from out of the whole complex of facts must be made by applying common-sense standards. Causation is to be understood as the man in the street, and not as either the scientist or the metaphysician, would understand it."[109] To sum up, proximate cause is identified by effect and not by timing. The meaning of "arising from" is similar to "resulted from". The latter can be also explained by a "proximate cause" test. Under the basic principle, where the loss has multiple proximate causes and one of which is specifically excluded by the policy, the exclusion prevails and the underwriters are not liable.[110] Accordingly, where pollution damage is caused by multiple proximate causes, one of which falls within the terms of the exceptions contained in Article 3(3)(a), no liability shall attach to the shipowner. In other words, the shipowner can be exonerated from liability if the "act of war, etc." is the "proximate cause" of oil pollution. In addition, if there are concurrent causes for the pollution accident, it is important for the shipowner to prove that the "act of war, etc." and other causes are the proximate causes concurrently in order to exonerate himself from liability.

The use of the word "wholly" in the relevant sub-paragraphs is very significant in that it substantially restricts the scope of applicability of the exception.[111] It also makes clear how to ascertain the circumstances in which the shipowner will be exonerated from liability. The shipowner can be relieved of his liability to pay

---

104 For more details read Bennett, Howard, *supra*, note 92, in: Chapter 6, pp. 115-135.

105 Templeman, Frederick, *Templeman on Marine Insurance: its Principles and Practice* (1986), 6th ed., pp. 190-210; see also Ivamy, E R Hardy: *General Principles of Insurance Law* (1993), 6th ed., pp. 406-420.

106 [1918] A.C.350.

107 *Ibid.*, at 369.

108 *Yorkshire Dale Steamship Co. Ltd. v. Minister of War Transport (The 'Coxwold')* H.L. (1942) 73 Ll.L.Rep. 1.

109 *Ibid.*, at 10.

110 Bennett, Howard, *supra*, note 92, p. 447.

111 Gauci, Gotthard, *supra*, note 5, p. 75.

compensation if he can prove the intention of a third party or negligence or other wrongful act of government. In a bunker-oil spill incident, the enumeration of the causes is exclusive. If there are other contributory causes, the provision is not applicable.

The inclusion of "resulted wholly or partially from" in Article 3(4) implies, on the other hand, a contributory liability. The wording of "resulted wholly or partially from" means that it is sufficient that the fault of the victim is a sole cause or a contributory cause of pollution damage and that the condition of being "wholly or partially" is very much dependent on the facts or according to the extent of the fault of the victim.

In brief, strict liability is strict in essence, but it also has exceptions. The application of those exceptions depends on the facts of the case.

III. Channelling of liability

Channelling of liability is an important and sensitive issue. It is important, because it links the issue of strict liability and the issue of insurance; it is sensitive, because once a party or person is chosen as the one to whom all the liabilities will be channelled, he will be the first to assume a potentially huge liability burden in an oil-pollution incident. It is therefore relevant to ask: to whom should the liability be channelled? Or, is there a need to maintain the mechanism of channelling liability?

Under the CLCs, the channelling of liability is stipulated as follows: the liability is imposed on a party and no claim for compensation may be made against other parties involved in the same incident unless the damage resulted from these other parties' personal act or omission committed with intent to cause such damage, or done recklessly and with knowledge that such damage would probably result. Therefore, no other parties may be held liable towards victims, even though a general case of liability would *prima facie* seem to apply. By contrast, the Bunkers Convention does not contain a provision of "channelling of liability".[112] The absence of effective provisions regarding the channelling of liability may cause confusion to and difficulty for victims in determining the proper party against whom claims should be brought.

The concept of shipowner includes a group of liable persons under the Bunkers Convention;[113] it is possible for any of them to try to shift liability to others. Meanwhile, although victims can initiate proceedings against any liable person, the compensation from persons other than the registered owner may, however, not be guaranteed in the absence of insurance. Furthermore, the joint and several liability provision in the absence of channelling may also invite claimants to sue a range of parties.[114] It is likely that, in practice, disputes will arise concerning the

[112] See Chapter 1, Section C.IV.2.

[113] The Bunkers Convention, Art. 1(3).

[114] The Bunkers Convention, Art. 3(2): "Where more than one person is liable in accordance with paragraph 1, their liability shall be joint and several."

apportionment of the liability between or among different persons. This will delay the response to the incident.

By virtue of Article 3(5),[115] the shipowner can have complete immunity outside the Convention; in other words, it is impossible for victims to bring claims in respect of pollution damage against the shipowner outside the Bunkers Convention. The said provision does not, however, offer any protection to any other parties, such as salvors, the servants of the owner, or the members of the crew and so on. Therefore, it is possible that if an incident involves claims against those parties, the claims will have to be brought under national law, while claims against the owner will be brought under the Bunkers Convention.

By removing the channelling of liability provision, the protection to the salvors or persons taking measures to prevent or minimise the effect of oil pollution will also be removed. The Bunkers Convention might thus discourage prompt response to a bunker pollution incident, because the salvors might be worried about finding themselves liable for pollution damage caused by their salvage efforts. This so-called "responder immunity" issue was already debated at meetings of the IMO Legal Committee and was ultimately rejected. It would, however, be important to maintain a "responder immunity" which could encourage prompt and effective response and thereby minimise pollution damage. A resolution, aimed mainly to protect salvors, was adopted as demanded by salvors' interests.[116] In accordance with this resolution all State Parties implementing the Bunkers Convention may adopt their own domestic legislation with regard to the protection of salvors and other responders. For instance, Australia and the UK have given quite firm indications that their domestic legislation implementing the Bunkers Convention will contain "responder immunity" provisions.[117]

For practical reasons, the proper channelling provisions should have been included. But the warning was also given during the discussions of a bunkers convention that careful consideration had to be given in exempting any persons from liability since no second-tier remedy would be available.[118] In the end, the channelling provision was not included in the Bunkers Convention. However, it might be understood that the Bunkers Convention was intended to channel the whole liability to a group of liable persons. All parties – such as the registered owner of the vessel, the bareboat charterer, the manager and the operator of the ship – are liable for pollution damage. It thus exposes a large group of liable persons, who may consider taking out insurance, to pollution claims. However, the requirement in Article 7(1) on the registered owner to insure his potential liability may result in claims being channelled to the registered owner.

---

[115] The Bunkers Convention, Art.3(5): "No claim for compensation for pollution damage shall be made against the shipowner otherwise than in accordance with this Convention."

[116] See Chapter 1, Section C. VIII.

[117] Wu, Chao, 'Liability and Compensation for Bunker Pollution', 33 *J. Mar. L. & Com.* 553 (2002), pp. 553-567, at 560.

[118] See IMO LEG 75/ WP.1.

E. Implementation of liability: insurance

In practice, liability for oil-pollution damage involves a large amount of compensation. A single person is hardly able to pay such sums. It is thus important that the liable person can spread the risks through the medium of insurance. The insurance is principally concerned with mitigating the adverse consequence and is designed to provide cover in respect of fortuitous events such as an incident. Therefore, it is often the case that the party who may be held liable voluntarily purchases insurance. The compulsory insurance requirement of the conventions not only requires the liable party to take out insurance, but has also virtually affected the allocation of liability and relevant provisions.

The shipowner is held liable for pollution damage once the claimant shows the fact and the cause of the pollution damage. In the CLCs, the meaning of "owner" used in the liability provision is the same as that in the provision regarding compulsory insurance. They refer to "the person or persons registered as the owner of the ship or, in the absence of registration, the person or persons owning the ship. However in the case of a ship owned by a State and operated by a company which in that State is registered as the ship's operator, 'owner' shall mean such company."[119] In contrast, the Bunkers Convention holds a "registered owner, bareboat charterer, manager and operator of the ship"[120] liable for the pollution,[121] while only requiring the registered owner to take out insurance. It leaves other liable parties outside the compulsory insurance requirement. The person liable for purchasing insurance and the person liable for pollution damage may differ.

This is an unequal system in the sense that the registered owner and his liability insurer may be in effect expected to respond in the first instance with regard to the fault of another party. Moreover, in the absence of other liable parties' liability insurance companies or financial security institutions, it is doubtful whether the registered owner can successfully carry out the right of recourse against other liable parties.

I. Proposals during the preparatory work of the Convention

During the discussions, there was an alternative draft proposed by the Japanese delegation for the "compulsory insurance" provision.[122] It advocated that the insurance requirement should be placed totally on the "shipowner", i.e. not on the "registered owner" alone. The proposal read:

> "The shipowner of a ship having a gross tonnage greater than [...] registered in a State Party shall ensure, independently or jointly, as co-assured or a beneficiary, as the case

[119] The 1969 CLC, Art. I(3).
[120] The Bunkers Convention, Art. 1(3).
[121] The Bunkers Convention, Art. 1(3).
[122] See IMO LEG 81/4.

may be, that there is in place insurance or other financial security, such as a bank guarantee, to cover the liability of the shipowner under this Convention."[123]

Since every individual included in the definition of shipowner was required to be liable for bunker pollution damage, the compulsory insurance requirement should also be extended to all members of that group.[124] This proposal was intended to ensure that all parties within the definition of "shipowner" were financially capable and that they would have sufficient financial security to cover the liability.[125] It was understood that the liability insurance taken out by the registered owner would not necessarily cover the liability of other parties who might be liable under the bunkers convention.[126] Furthermore, in the absence of a fund similar to the IOPC Fund, ratifying States might require further reassurance that the recovery would be possible, in case the security of the registered owner failed.[127] However, this proposal was not accepted.

The Republic of Korea also submitted a proposal in this respect. The said proposal retained the coverage of a small group defined as the "shipowner" for the purpose of liability, but would require the insurance only to cover the liability of the registered owner and not multiple insurance policies.[128] This modification was founded on two reasons. The first was that a joint and several liability mechanism would be provided in the convention. In general terms, a "joint and several liability" means the obligation of "two or more debtors who are obliged to the same creditor for the same obligation… where any of them may be compelled to perform the whole obligation and where performance by one of them releases the other debtors towards the creditor…"[129] Pursuant to this "joint and several liability" provision in the convention, the victims would be given the possibility of claiming full redress from any party responsible for the damage. Other parties were to be liable even if they had not established any financial security. The second reason was that there would be a right of recourse for the registered owner.[130] Therefore, there would be no gap in the liability regime.[131] This proposal received a significant amount of support; but was still not accepted.

---

123 See IMO LEG 81/4.

124 See IMO LEG 81/WP. 2.

125 See IMO LEG 81/WP. 2.

126 See IMO LEG/CONF.12/9, submitted by the International Group of P&I Club.

127 See IMO LEG/CONF.12/9.

128 See IMO LEG 80/4/1.

129 This definition is from William Tetley, Glossary of Maritime Law Terms (2004), available at: <http://www.mcgill.ca/maritimelaw/glossaries/maritime/#letter_j> (visited 1 June 2005).

130 The Bunkers Convention, Art.3(6): "Nothing in this Convention shall prejudice any right of recourse of the shipowner which exists independently of this Convention."

131 See IMO LEG 81/WP.2.

II. Two alternative means

Although the above-mentioned two proposals were not accepted during the drafting work of the bunkers convention, two questions should be considered in more detail: first, whether those potentially liable parties can be required to take out insurance separately; and secondly, whether the parties involved, for example, the shipowner and the charterer, can become the co-assured in one insurance policy. The following sections will exam these two possibilities.

1. *Separate insurance policies*

In theory, if the Bunkers Convention required all liable parties involved to take out insurance, it would increase the chances for claimants to obtain adequate compensation after a bunker spill incident. However, the disadvantages would be obvious.

If all parties were required to take out insurance, more insurers would be involved. In practice, different liability insurers insure different parties engaged in the same venture.[132] Accordingly, on the one hand, this would impose a higher burden on the insurance industry and its administration; on the other hand, once a bunker spill incident occurred, the several parties and their insurers would be unlikely to agree quickly on the practical apportionment of liability, which would consequently retard the speed of response. The claimants might also have difficulty in identifying the liable party and initiate unnecessary litigation. In the end, it would not improve the situation of the claimants; it would, however, complicate the case and confuse claimants.

It may be instructive to compare two cases: *Rosebay*[133] and *American Trader*[134], which both involved similar amounts of oil cargo spilled, one in the United Kingdom, which is a member of the CLCs, and the other in the United States, which has its own OPA 90. The *Rosebay* was settled satisfactorily within two years, whereas the *American Trader* was resolved nearly ten years later. One of the reasons for the delay in the latter case was the multiplicity of parties sued.[135] Under the OPA 90, the responsible party, in the case of a vessel, means any person owning, operating or demise chartering the vessel. All of them are obliged to establish or maintain evidence of financial responsibility sufficient to meet their

---

[132] See IMO LEG/CONF.12/9, Submitted by the International Group of P&I Club.

[133] After a collision with the trawler Diane Marie on 12 May 1990, the tanker *Rosebay*, sailing from the Persian Gulf to Rotterdam, spilled 1,073 tons of crude oil in the open sea. Oil slicks were threatening the beaches of south Devon. Response operations were set up immediately. As a result, only a small amount of mousse reached the coast.

[134] The *American Trade* oil spill happened in 1990 off the shores of Huntington Beach in Orange County, California. It spilled approximately 400,000 gallons of Alaskan North Slope crude oil. Although a US$3 million settlement was agreed upon shortly after the spill, due to challenges to the settlement from non-settling defendants, court challenges delayed the availability of funds until 1998.

[135] See IMO LEG 80/4/2.

liability under the OPA 90.[136] Such provisions can prompt victims to claim against all possible responsible persons.

In practice, the identification of the registered owner is relatively easy, since it is based on the administrative evidence – the person or persons registered as the owner of the ship or, in the absence of registration, person or persons owning the ship.[137] The registered owner is thus required to take out insurance. In effect, even if all parties were required to take out insurance, it is most likely that victims would prefer to file a claim against the most easily identifiable person, i.e. the registered owner.

2. *Co-assurance under one policy*

Another question is whether it would be feasible to let all parties agree to take out insurance as co-assured parties under one policy, i.e. all liable persons involved will be insured in one insurance policy against oil-pollution liability. During the discussions of the bunkers convention, the International Group of P&I Clubs expressed their apprehension that it would be difficult for the Clubs to provide comprehensive insurance cover for a vessel when all persons that was to be defined as the owner would be jointly and severally liable, especially when the Club did not even know some of those persons. Such an arrangement would provide difficulties for the mutual arrangement on which the Clubs were based.[138]

In practice, a P&I Club membership is usually made up of shipowners, corporate and individual, managing owners, ships' operators and charterers.[139] Most members of a Club are shipowners. A charterer may become a member in a Club as a "special entry". However, the shipowners are usually not willing to offer benefits to other groups of persons.

The charterers may join the owner's Club separately or under a "family arrangement", whereby both the shipowner and the charterer enter the same Club as co-assured parties and co-members.[140] If the registered owner and the charterer agree to take out P&I insurance for bunker-oil pollution liability together, one may ask whether this arrangement implies a better compensation situation for claimants. In the case of joint entries, whereby the co-assured parties and affiliated members are covered under group affiliate arrangements for so-called "misdirected arrow" claims, the Club Rules provide that the Club's liability will not exceed the limit of liability of the shipowner-member. In other words, if the shipowner and his charterer are insured under one P&I policy, the amount of the insurance available for pollution victims is the same as if the shipowner were insured alone. Therefore, the situation of the claimant would be unchanged.[141]

---

136 The OPA 90, §§ 2701(32)(A).
137 The Bunkers Convention, Art. 1(4).
138 See IMO LEG 80/4/1.
139 Hazelwood, Steven J., *P& I Clubs: Law and Practice* (2000), p. 83.
140 *Ibid.*, p. 99.
141 *Ibid.*, p. 385. According to Club practice, if the owner's insurance and charterer's insurance are covered with the same Club for one and the same vessel and both

It is possible that liability insurers other than P&I Clubs intend to insure oil-pollution liability. However, difficulties may arise in those cases. Apparently, it is rare for one insurer to cover all different liable persons that are nominated in the Bunkers Convention. Additionally, another practical difficulty is that charterers or managers often charter or manage hundreds of ships, and the insurance for those ships is often placed on a block basis so as to take advantage of the distribution of risk within the block. This is plainly a different spreading of risks from that offered by the owner's fleet.[142] Therefore, if an incident occurs to one particular ship, it is not easy for the charterer or the manager to arrange the insurance indemnification for that ship.

From the preceding explanation, it is clear that neither separate insurance policies nor co-assurance under one policy may gain much support.

F. Concluding remarks

Strict liability is the chosen liability rule in the Bunkers Convention, and insurance is also required. In practice, even where there is no legal duty to insure against liability, the liable person tends to voluntarily procure the insurance policy against such a large-scale liability.

In earlier conventions, the person who can most easily be identified was chosen as the liable person and also the person to take out insurance. The answer to the questions of who shall be liable and who is in a better position to insure against liability under the Bunkers Convention is, however, novel to other civil liability conventions: the shipowner that includes a group of different persons is strictly liable for pollution damage; however, only the registered owner is required to take out insurance.

members become liable, the Club's own retention will be twice U.S. $5 million, which was the basic retention for the member at the time.

142 See IMO LEG 80/4/2.

Chapter 4: Insurance and the Quest for Adequate Compensation

A. Introduction

The Bunkers Convention was adopted to ensure the payment of adequate, prompt and effective compensation for the damage caused by pollution resulting from the escape or discharge of bunker oil from ships.[1] Compensation is a common form of reparation where restitution in kind is impossible.[2] Starting with the 1969 CLC, the maintenance of insurance or other financial security has been designed for the purpose of compensation.[3] The liable person shall take out liability insurance or choose to establish other types of financial security to meet pollution victims' claims. Additionally, it follows that the soundness of "other financial security" must be similar to that of a bank.[4]

The Bunkers Convention underlines the importance of liability insurance and financial security. One of its main points is that the security should be financially adequate. Due to the lack of a second-tier compensation source, it is relevant to examine: first, is the requirement regarding *compulsory insurance or financial security* under the Convention an effective and adequate safeguard for compensation? Secondly, is it feasible to arrange other compensation sources? This chapter will attempt to focus on these two issues.

B. Certification of insurance

I. Basic requirements in the Bunkers Convention

There is always a list of certificates and documents which should be carried on board ships [5] for the purpose of complying with international or national requirements. At the beginning, these documents were mainly related to the safety

---

1 See the Preamble of the Bunkers Convention.

2 Springer, Allen L., *The International Law of Pollution: Protecting the Global Environment in a World of Sovereign States* (1983), p. 136.

3 1969 CLC, Art. VII.

4 The Bunkers Convention, Art.7(1).

5 For example, see IMO Ref. T3/2.01 – Revised list of certificates and documents required to be carried on board ships.

of the ship. Later on, additional documents were also required for other aspects regarding such as compliance with environmental standards.[6]

With the advent of the CLCs, each ship that is subject to the conventional requirement must take out insurance to meet its oil-pollution liability. A certificate issued by the relevant authority must be carried on board. This certificate attests that insurance or some other financial security for pollution liability is in place. In line with the CLC model, the Bunkers Convention also contains provisions requiring a ship to have on board a certificate confirming that insurance or other financial security actually exists.[7] The form of the certificate is specified and the certificate itself, denominated as "Certificate of insurance or other financial security in respect of civil liability for bunker oil pollution damage" as set out in the Annex to the Convention, is subject to inspection.[8] The appropriate authority of the State of the ship's registry in a State Party is responsible for the issuance of a certificate confirming that the appropriate insurance or financial security is in place.[9] The international recognition of certificates and the maintaining of certificates in electronic format are also provided for in Article 7 of the Bunkers Convention.[10]

II. Administrative burden corollary to the issuance of the certificate

Fears were expressed during the discussions of a bunkers convention that the arrangement of compulsory insurance would be administratively burdensome, since the insurance or other financial security has to be certified.

Although the CLCs only impose an insurance requirement on oil tankers carrying certain tons of oil, it is already very cumbersome to satisfy the CLC certificate requirements. As indicated by one author:

> "In that system, each vessel needs a paper certificate on board, which must be renewed regularly. Each renewal involves the P&I club, that issues a so-called blue card, the governments, that issue or authorize the certificate and scrutinize the insurer, and finally a logistics problem in getting the certificate on board the vessel in time."[11]

Apparently, the procedure for the processing of the certificate used in the CLCs involves extensive administrative commitments.

---

6 Molenaar, Erik Jaap, *Coastal State Jurisdiction over Vessel-Source Pollution* (1998), p. 36.

7 The Bunkers Convention, Art. 7(2) and Art. 7(4).

8 The Bunkers Convention, Art. 7(2).

9 The Bunkers Convention, Art.7(7). Art. 7(2) provides further that, if a ship is not registered in a State Party to the Bunkers Convention, the certificate has to be issued by the appropriate authority of any State Party.

10 For in more details please see Chapter 1, Section C.V.2.

11 Rosag, Erik, 'Compulsory Marine Insurance', originally published in *Scandinavian Institute of Maritime Law Yearbook 2000*, available at: <http://folk.uio.no/erikro/WWW/corrgr/insurance/simply.pdf> (visited 21 March 2005).

The Bunkers Convention, once ratified, will involve a much larger number of ships of different types, thus intensifying the administrative burden.[12] In 2002, the inspections by the UK Club taken in some countries[13] indicated that about one third of inspection visits were made to bulk carriers, around 18% to tankers, nearly 16% to container ships and 9% to general cargo vessels.[14] This, to some extent, showed that around 80% of the vessels would come within the requirements of the Bunkers Convention if it were in force, although it did not necessarily mean that all or most of them would need to take out insurance.

Under the Bunkers Convention, the certification of insurance essentially shows the implementation of compulsory insurance. If this does not work out properly, the requirement of compulsory insurance may lose some value. Different views and suggestions on alleviating the administrative burden were advanced at the IMO Conference. The following surveys will be focused on the flag States and the port States. One pertains to the issuance of certificates of the flag States, and the other relates to the role of the port States in terms of control.

1. The administrative burden of the flag States

In general, a ship comes within the national jurisdiction of a State after the registration in that State. The State thereafter exercises the power inherent in the jurisdiction as being a flag State in assuming authority over the ship. The State also undertakes the national and international responsibilities of a flag State in relation to that ship.[15]

As far as the issue of the insurance certificate is concerned, by virtue of the relevant provisions in Article 7 of the Bunkers Convention, the responsibilities of a flag State if it is a State Party to the Bunkers Convention, are mainly: first, to establish or authorise an appropriate institution or organisation to issue certificates; second, to fully guarantee the completeness and accuracy of the certificate so issued; third, to notify the Secretary-General of the IMO of the specific arrangements of the authorised institution or organisation for issuing certificates.[16] According to the experience in issuing CLC certificates and the facts that more ships of different types will be involved under the Bunkers Convention, the above provisions will in effect entail arduous administrative procedures and paperwork on the part of the flag States.

It will be necessary to have the insurance certificate in place, since it will be the evidence showing the financial security of the liable person. Due to the administrative burden possibly arising, some alternatives were considered in order to alle-

---

12 See IMO LEG 76/WP. 3 Agenda item 4; also see IMO LEG 77/6 and IMO LEG/CONF.12/10.

13 These include the Netherlands, the US, South Korea, China, Taiwan, Italy, India, Egypt, the United Arab Emirates and Singapore.

14 UK P&I Club, 'Port State Control', *P&I Int.*, 2003, 17(5), at 14.

15 Ozcayir, Z. Oya, *Port State Control* (2001), p. 10.

16 The Bunkers Convention, Art.7(2)-Art.7(7).

viate the envisaged administrative burden of a flag State. The following sections will assess these alternatives:

Alternative 1: Certificate issued under the IMO Assembly resolution A.898 (21)

During the IMO Legal Committee's conferences for the proposed bunkers convention, the International Chamber of Shipping suggested that the requirement of State-approved certificates should be abandoned. Instead, the proposal was that Article 7 of the Bunkers Convention should provide that a State Party "may require the registered owner to demonstrate evidence of financial security in accordance with the international standards adopted by the Organisation".[17] The so-called "international standard" is based on the IMO Assembly resolution A.898(21) – "Guidelines on Shipowners' Responsibilities in Respect of Maritime Claims".[18] Article 5.1 of the Guidelines provides that,

> "Shipowners should ensure that their ships have on board a certificate issued by its insurer. Where more than one insurer provides cover for relevant claims, a single certificate confirming the identity of the main liability insurer is sufficient."

The "insurer" means "any person providing insurance for a shipowner."[19] "Insurance" in these Guidelines means "insurance with or without deductibles, and comprises, for example, indemnity insurance of the type currently provided by members of the International Group of P&I Clubs, and other effective forms of insurance (including self-insurance) and financial security offering similar conditions of cover."[20]

If the Bunkers Convention agreed to the registered owner demonstrating evidence of financial security in accordance with the Guidelines, the shipowner himself would arrange insurance cover for his ship to comply with the Guidelines. The insurance would respond up to certain limit such as the limit set under Articles 6 and 7 of the LLMC.[21] The certificate would be issued by the insurer.[22]

Accordingly, this arrangement would mean that the flag States would not need to set up or authorise the appropriate institutions or organisations for the required State-proved insurance certificates under the Bunkers Convention. It would thus alleviate substantially or even remove the administrative burden from the flag States. However, this alternative, if adopted, would increase the administrative burden of the port State. The port States might have to pay special attention to checking whether an entering ship had included bunker-oil pollution liability in its insurance coverage as required under the Bunkers Convention. They would have to evaluate the certificate individually in some extreme cases.

---

17 See IMO LEG/CONF.12/10.

18 In practice, this would mean that ships should carry on board P&I Club "Certificates of Entry" (or other evidence of insurance) as recommended in this IMO resolution: see IMO LEG/CONF.12/10.

19 See IMO A21/Res.898, available at: <http://folk.uio.no/erikro/WWW/ corrgr/insurance/898.pdf> (visited 5 June 2005).

20 See *ibid.*

21 *Ibid.*, Art.4.1.

22 *Ibid.*, Art.5.1.

Alternative 2: Authorising the International Group of the P&I Clubs to issue the certificate

The MARPOL 73/78 Conventions and the SOLAS (the International Convention for the Safety of Life at Sea)[23] have a mechanism for authorising classification societies to issue survey and safety certificates on behalf of the national administration. The delegation from Australia noted that this approach could be adopted for the purpose of the bunkers convention with modifications.[24] The basis of such an approach was to be founded in the IMO Assembly Resolution A.739(18), namely the "Guidelines for the authorization of organizations acting on behalf of the Administration". This resolution was adopted on 4 November 1993 and aimed to develop uniform procedures and a mechanism for the delegated authority, as well as the minimum standards for recognised organisations acting on behalf of the Administration. The enforcement of the Resolution was to assist the flag State in the uniform and effective implementation of the relevant IMO conventions.[25]

In Australia, only the classification societies which were full members of the International Association of Classification Societies were authorised.[26] It was therefore considered that the draft bunkers convention could be amended to allow similar arrangements to be made for insurance providers.[27] This meant that national administration agencies concerned would be able to authorise special insurance providers to issue the required certificate according to the requirements of the bunkers convention. In practice, the members of the International Group of P&I Clubs might be authorised to issue the required certificate.[28] The main advantage of this alternative was that it would alleviate the burden put on the State Party to establish or authorise the appropriate institution or organisation to issue the certificate.

Some delegations, however, felt that this approach posed numerous difficulties. First, it was not possible to compare insurance providers with the classification societies. Besides, it would not be acceptable to entail the duties of a delegating State to private concerns. Secondly, it would involve a selection or preference of one company over other companies, which would run counter to the WTO, EC and some national competition laws.[29] However, if this arrangement apparently infringed WTO, EC and some other national competition rules, the above-mentioned arrangements in relation to classification societies would thus also contravene such rules, yet they exist in practice.[30] Most importantly, the representatives of P&I Clubs in the negations expressed the view that an analogy to the

---

23 The SOLAS are reprinted in: *Lloyd's Shipping Law Library: The Ratification of Maritime Conventions* (2004), Vol.2, II.3.10, II.3.20, II.3.21, II.3.22, II.30, II.3.40, and II.3.41.

24 See IMO LEG 76/WP. 3, agenda item 4.

25 See IMO Resolution A.739(18).

26 See IMO LEG 77/6.

27 See *supra*, note 24.

28 See *supra*, note 26.

29 See *supra*, note 24.

30 See *supra*, note 26.

certification by classification societies was inappropriate in the context of the certificate of P&I insurance. It was indicated that a regime dealing with oil tankers was very different from a regime dealing with all dry cargo vessels in respect of the number, trade and effective port State control. The conclusion was that this would be unworkable.[31] Therefore, in the end, the alternative of authorising the International Group of P&I Clubs to issue the certificate was rejected.

Alternative 3: P&I "Certificate of Entry" replaces insurance certificate under the Convention

The "Certificate of Entry" of a P&I Club is the official record attesting that the request of a ship to be entered in a Club has been accepted. This Certificate can be carried on board and used as the evidence of P&I insurance cover when required under international or national legislation. The above-discussed *Alternative 1* shows that the certificate issued under IMO Assembly resolution A.898(21) would to great extent mean that ships would carry on board the "Certificate of Entry" from a P&I Club.[32] However, what is embodied in that alternative would also include other effective forms of insurance. Additionally, not only P&I Clubs in the International Group of P&I Clubs, but also many other P&I Clubs[33] and other liability insurers can offer liability insurance for bunker-oil pollution risk. Other liability insurers or P&I Clubs outside the International Group cover around 5%-10% of worldwide shipping tonnage.[34] Hence, we have a separate discussion here.

If a bunkers convention allowed the registered owner to carry on board the "Certificate of Entry" of any P&I Club, it would mean that regulations were needed to ascertain the financial strength contained in the certificate. At the time of the IMO conference, it was indicated that only the "Certificate of Entry" issued by a reputable P&I Club could itself be relied on to show that there was adequate cover for bunker spills in most cases.[35] However, the preference of one P&I Club, and even other Clubs, would run counter to the WTO, EC and some other national competition rules.

If this alternative were adopted, it would also need cooperation from P&I Clubs. First, the Clubs should maintain the liability coverage for bunker pollution risks. Secondly, in P&I practice, a "Certificate of Entry" is not regarded as conclusive with regard to the risks which are endorsed on it. For during the period of membership, there may have been a variation in insurance coverage, which, although possibly recorded on a so-called "Endorsement Slip", will not appear on the original "Certificate of Entry".[36] Thus, if the bunkers convention had allowed the registered owner to carry on board the "Certificate of Entry" instead of a specific certificate for bunker-oil pollution, it would have introduced some

---

31 See IMO LEG 77/11.

32 See IMO LEG/CONF: 12/10.

33 These other P&I Clubs mainly mean the Clubs outside the International Group of P&I Clubs.

34 This percentage is inferred from the fact that more than 90% ships by tonnage are covered by the Clubs of the International Group of the P&I Clubs.

35 See IMO LEG 80/4/2.

36 Hazelwood, Steven J., *P&I Clubs Law and Practice* (2000), p. 43.

uncertain factors regarding insurance coverage. The port States would have to verify membership details with the Club's managers all the time to ascertain a member's precise scope of cover.

A similar approach is, nonetheless, utilised in practice. From March 1, 2005, the shipowner of a vessel calling at any Japanese port must have insurance for any vessel of 100 gross tons or more. If the shipowner has taken out insurance from "designated insurers", such as P&I Clubs belonging to the International Group of P&I Clubs, the original or the certified copy of the "Certificate of Entry" issued by the insurer will be accepted as a substitute for the certificate of insurance issued by the Minister of Land, Infrastructure and Transport.[37]

Alternative 4: Minimising the coverage of the Bunkers Convention

The Bunkers Convention will impose a heavy administrative burden on the flag States to issue insurance certificates because it aims to cover all types of ships. There were conflicting interests relating to the insurance threshold and the coverage of the ships during the discussion of the convention. On the one hand, the proposed convention was expected to cover as many ships as possible in order to promote a more comprehensive protection and compensation for oil pollution risks. On the other hand, the capacity of the insurance market as well as the administrative burden involved therein could not be overlooked. These opposing concerns, therefore, led to protracted discussions, some of which have already been elaborated in Chapter 1.[38] In the provision finally adopted in the Convention, not all ships are required to take out insurance. Only those having a gross tonnage greater than 1,000 and registered in a State Party are required to do so. In the discussions on the basic standards for the insurance requirement, it was recognised that a "*400 gross tonnage insurance threshold*" is preferable.[39] However, in the end, the compromise figure of 1,000 gross tons was preferred and adopted by the Convention.

Unfortunately, the result adopted cannot cover all ships with the potential to cause pollution damage, since serious pollution damage can be also done by a ship with a gross tonnage below the figure of 1,000. Although a large number of ships still fall within the compulsory insurance requirement, the provision in the Convention has relatively reduced the number of ships that will be required to take out insurance when the Convention takes effect. In addition, it must be admitted that this provision will, to some extent, reduce the administrative burden.

As seen, the last alternative has been written into the Convention. Through the setting-up of a comparatively high insurance threshold, the administrative burden will probably be eased. Meanwhile, compulsory insurance or financial security and the State-proved certificates have been chosen as the most manageable option.

---

37 Japan, Minster of Land, Infrastructure and Transport: 'Introduction of Compulsory Insurance Requirement for Non-tanker Ships', available at: <http://www.mlit.go.jp /english/maritime/Web/insurance_portal3eng.htm> (visited 17 November 2005).

38 See Chapter 1, Section C.V.1.b).

39 See IMO LEG/CONF.12/6. For more details about the "insurance threshold", see Chapter 1, Section 3.5.1(2). It most closely reflects current practice in international conventions in which a starting point of ship's gross tonnage is required to be set.

2. *Port State control regarding the certificate*

According to the Bunkers Convention, if a port State is a State Party to the Bunkers Convention, it will have similar tasks as listed in the above section regarding the responsibilities of a flag State, such as the authorisation of an institution or an organisation to issue the certificate.[40] Besides, the activities involved in "port State control" include the idea that a port State conducts the inspection of foreign ships in its ports and finds out whether those foreign ships comply with internationally required standards. Therefore, among the usual routine things, the port State shall ensure that insurance or other financial security is in force and complies with the requirements of the Bunkers Convention.[41]

Ships are required to carry on board certificates in accordance with many IMO conventions. The port State authority shall already carry out inspections of numerous certificates carried by a ship.[42] As for the ship, any failure to present the required certificates may result in detention of the vessel in a port. Given that such a system already existed, it was hoped that the administrative burden imposed by the processing and inspection of the insurance certificates under the Bunkers Convention should not be so onerous.[43]

During the meetings of the IMO Legal Committee, it was pointed out by Australia that the States already had various arrangements in place to check the validity of certificates carried on board. In practice, more than 25% of all non-tankers over recent years have been inspected by port state control inspectors on arrival in Australia. Therefore, "...The addition of one more certificate to be checked is at worst a minimal increase in required effort (given that this check is already required for tankers) compared to the potential advantage to the States, especially to the coastal State."[44] The implicated administrative burden is deemed to be minimal as compared to the advantages it will bring to the coastal State, such as having guaranteed compensation available once the incident occurs in its port area.

III. Electronic means for the certificate

The insurance certificate is usually a paper certificate, but the Bunkers Convention also allows a certificate to be "in an electronic format". Article 7(13) provides that:

> "Notwithstanding the provision of paragraph 5, a State Party may notify the Secretary-General that, for the purposes of paragraph 12, ships are not required to carry on board or to produce the certificate required by paragraph 2, when entering or leaving ports or arriving at or leaving from offshore facilities in its territory, provided that the State Party which issues the certificate required by paragraph 2 has notified the Secretary-General

40 The Bunkers Convention, Art.7(3).
41 The Bunkers Convention, Art. 7(12).
42 Ozcayir, Z.Oya, *supra*, note 15, pp. 102-105.
43 See IMO LEG 77/11.
44 See IMO LEG 77/6.

that it maintains records in an electronic format accessible to all State Parties, attesting the existence of the certificate and enabling State Parties to discharge their obligations under paragraph 12."[45]

In brief, the electronic certificate may substitute the paper certificate.

The acceptability of the certificate in an electronic format is a result of technical development and practical need. In many cases, the traditional way of carrying transport information, i.e. on a piece of paper, has been replaced by a record in a computer, which acts as the carrier of the information. Computer-based operations may also minimise administrative tasks. Similar methods have already been applied in many countries. For instance, they use radio customs clearance, prior to the arrival of the vessel, to deal with most required certificates.

At the time of the discussion of a convention for bunker-oil pollution liability, it was suggested that the checking and clearance of the certificates should be done electronically by the customs authorities as one part of port State control, since a large number of certificates would be involved. In that way, it would reduce the extra administrative burden involved in issuing or inspecting the certificates. A number of delegates spoke in favour of this proposal.[46] However, the International Group of P&I Clubs worried about the effect of this proposal. They argued that port State control had been introduced because of the inability of many flag States to ensure compliance with IMO Conventions. Additionally, it was not really convincing that flag States would be capable of maintaining relevant records attesting the existence of the insurance certificates in an electronic format, which would involve a lot of financial and technical problems. Moreover, even if this were proved to be possible in practice, it would be doubtful whether the administrative burden would be much reduced.[47]

Nevertheless, the admission of the certificate in an electronic format was adopted in the Bunkers Convention. If a State Party decides to maintain the electronic means for insurance certificate, it will at least involve the cumbersome work of putting relevant data into computer databases.

IV. The validity of the certificate

To maintain the validity of the certificate, most of the formal requirements, including the language used, have been discussed in a previous section.[48] Caution required that more provisions should be written into the Bunkers Convention. Therefore, it follows from some other provisions that the insurance arrangement or other financial security will not satisfy the requirements if they can be terminated for reasons other than the expiry of the period of validity.[49] Even in the normal case, the notice of termination should be given three months before the termina-

---

45 The Bunkers Convention, Art. 7(13).

46 See IMO LEG 76/WP. 3, Agenda item 4.

47 See IMO LEG/CONF.12/9.

48 Section B.I of this chapter.

49 The Bunkers Convention, Art. 7(6).

tion.[50] The institution or organisation authorised to issue certificates in a State Party shall, as a minimum, be authorised to withdraw these certificates if the conditions under which the certificates have been issued are no longer maintained.[51] In all cases, the institution or organisation shall report such a withdrawal to the State on whose behalf the certificate was issued.[52]

The issued certificate shall attest that insurance or other financial security is in force for the purpose of the Bunkers Convention. Article 7(2)(f) specifically provides that the "...period of validity of the certificate which shall not be longer than the period of validity of the insurance or other security." More essentially, once liability has been established, the certificate must be valid to ensure the accessibility of compensation to eligible claimants. It is important to ensure to claimants at the time and place of an incident that: (1) the certificate concerned was issued by the authorised organisation of any State Party to the Convention; (2) the certificate embodies a guarantee of sufficient financial strength; and (3) the certificate can be valid at the time and place of the incident.

Pursuant to Article 7(7), the State of the ship's registry shall determine the conditions of issue and validity of the certificate.[53] In addition, it is clear from Article 7(2) that when a ship is registered in a State Party to the Bunkers Convention, a certificate must be obtained from the appropriate authority of the State of the ship's registry. A ship registered in a State that is not party to the Bunkers Convention will have to obtain the certificate from any State Party.[54] In the latter case, a certificate issued by the State of the ship's registry will not be recognised by the State Party to the Bunkers Convention. However, from the wording of Article 7(7), it is not clear whether any State Party that is not the State of the ship's registry and may intend to issue the certificate has the right to determine the conditions of issue and validity of the certificate. The answer might be affirmative according to Article 7(2), since it requires the State Party to determine that the requirements regarding the certificate under the Convention

50 The Bunkers Convention, Art. 7(6) and Art.7(5). The notice shall be given to the authority that keeps the record of the ship's registry or, if the ship is not registered in a State Party, to the authority issuing or certifying the certificate.

51 The Bunkers Convention, Art. 7(3)(c).

52 The Bunkers Convention, Art. 7(3)(c).

53 The Bunkers Convention, Art.7(7).

54 The Bunkers Convention, Art. 7(2). We should take a careful look at Art. 7(1) and Art. 7(2). Art.7(1) provides that there are two factors to be considered for deciding which registered owner should take out insurance: (1) the gross tonnage of the ship is greater than 1000; (2) this ship is registered in a State Party. The second factor means that if the ship is not registered in a State Party, it is not required to take out insurance even if this ship's gross tonnage exceeds 1000. However, under Art.7(2), it mentions, "with respect to a ship not registered in a State Party it may be issued or certified by the appropriate authority of any State Party." This can be understood as meaning that a ship that is not registered in a State Party should also purchase insurance. The certificate issued by a non-State Party is, however, not valid because the relevant certificate has to be issued by "any State Party".

have been complied with before issuing the certificate, but it would be much better if Article 7(7) could expressly provide that:

> "The State of the ship's registry shall, subject to the provisions of this Article, determine the conditions of issue and validity of the certificate. If the State of ship's registry is not a Party State to the Convention, any Party State intending to issue the certificate shall determine the conditions of issue and validity of the certificate."

Such a complex situation is also due to the provision in Article 7(11), under which a ship from a State that is not Party to the Bunkers Convention and wishing to enter the port of a Contracting State shall also carry on board a certificate.[55] It means that a ship, whether it is registered in a State Party or not, shall have the required certificate once it wishes to enter or leave the port of a State Party. However, in this case, it is hard to subject a State of the ship's registry, if it is not a Party State to the Bunkers Convention, to determine the conditions of issue and validity of the certificate or other relevant provisions in Article 7, since there may be few mutual interests between shipowners and their State of registry.[56]

Article 7(9) also provides that the certificate issued by any State Party shall be regarded as "having the same force as certificates issued or certified by them".[57] The certificate issued by one State Party shall be considered final by other State Parties.

C. Availability and capacity of insurance for bunker-oil spill liability

Given the fact that the registered owner is required to obtain compulsory insurance or other financial security, it is pertinent to consider "an adequate supply of suitable products"[58] and their capacity. Above all, except for the possibility that the shipowner may establish self-insurance, if the current market is not ready or is not willing to support and provide products, it will be pointless to require the shipowner to take out any financial security. There may be a shortage of available insurance;[59] most conventions thus "...require operators to establish some form of

---

55 The Bunkers Convention, Art.7(11): "A State Party shall not permit a ship under its flag to which this Article applies to operate at any time, unless a certificate has been issued under paragraph 2 or 14."

56 Bennett, Paul, 'Mutual Risk: P&I Insurance Clubs and Maritime Safety and Environmental Performance', *Marine Policy* 25 (2001), pp. 13-21, at 14.

57 The Bunkers Convention, Art.7(9).

58 Wilde, Mark, *Civil Liability for Environmental Damage* (2002), p. 296.

59 Pfennigstorf, W., 'Policy Considerations for Insurers Engaging in Environmental Liability Insurance', in: Bocken, H./Ryckbost, D. (eds.), *Insurance of Environmental Damage* (1991), pp. 269-289, p. 273, "...the availability of insurance coverage has become a factor of increasing importance in the development of the law of liability. This has become evident not only in decisions made by courts, but also in the process of legislation, especially with respect to environmental liability. There is strong reluctance to act without first obtaining a commitment from the

'financial security' for increased liabilities, whatever form that may take, rather than imposing a specific compulsory insurance component."[60] This provides the shipowners with "an element of choice".[61] Nevertheless, insurance is the main means of covering the liability.

The capacity of insurance is different from its availability. The availability of insurance may guarantee that the liability insurance for a particular type of risk is available. It may, however, occur that the available insurance cannot adequately supply the whole amount of required insurance, i.e. the capacity of insurance is insufficient. In practice, insurance capacity usually includes the capacity of direct insurance arrangements and reinsurance. Reinsurance is a traditional method for the insurer to protect himself against unexpected or excessive losses. It has been indicated that there would hardly be any liability insurance available at all for commercial risks without reinsurance.[62] This might be an exaggeration, but it is true that the insurer can pass on a part of the risk to the reinsurer through reinsurance, which to some extent alleviates the insurance burden and increases the capacity of insurance. However, the capacity of reinsurance is also limited, especially in response to some risks which may involve large-scale liabilities.[63]

It is clear that most of shipowners take out P&I insurance to meet their liabilities for oil-pollution risks. As far as P&I Clubs are concerned, they are not willing to cover oil pollution without limitation. They have consistently stressed that they are unable to provide financial security for unlimited amounts, and that a fixed limit is needed in any insurance certificate.[64] This relates to the general idea that in any system of compulsory insurance, an essential element is that there must be a limitation of the amount for which insurance must be maintained. Accordingly, the concept of "capacity of insurance" is transformed into "limit of insurance", although each is different from the other. In other words, even if the insurance industry is capable of satisfying liability insurance requirement without limitation, it prefers to impose a limitation on insurance coverage. More details about the interaction between insurance and the limitation of liability will be discussed in Chapter 5 of this research.

---

insurance industry to the effect that coverage commensurate to the intended new level of liability will be available."

60 Wilde, Mark, *supra*, note 57, p. 296.

61 Wu, Chao, *Pollution from the Carriage of Oil by Sea: Liability and Compensation* (1996), p. 53.

62 Pfennigstorf, W., 'Limited Insurability of Unlimited Liability: Serial Claims, Aggregates and Alternatives: The Continental View', in: Kroener, Ralph P(ed.), *Transnational Environmental Liability and Insurance*, pp. 159-165, at 161.

63 See 'The Underwriting of Oil Pollution Risks', in: DeLaRue, Colin M. (ed.), *Liability for Damage to the Marine Environment* (1993), pp. 149-154, at 152: "...the reinsurer might pay for the first and possibly the second oil pollution loss, but thereafter a company or syndicate writing the insurance would almost certainly be unprotected."

64 See IMO LEG 74/4/2.

D. Other related issues

Before discussing the effectiveness of this compensation system or the adequate payment it represents, three related issues need to be clarified. They are: (1) the "polluter pays" principle which is sometimes used to justify the strict liability rule under international civil liability conventions; (2) the financial standing of the insurer or guarantor, the fundamental importance of which relating to the compensation purpose of the current Convention deserves a separate discussion; and (3) the "mutuality", the feature central to shipowners' P&I Clubs, which basically ensures P&I Clubs' ability to insure large-scale third party liabilities.

I. "Polluter pays" principle

The "polluter pays" principle initially belonged to the domain of public law, since it was considered to be relevant to the cost of environmental protection measures, not *ex post facto* environmental harm.[65] This principle obliges the polluter to "bear the costs of measures to reduce pollution decided upon by public authorities to ensure that the environment is in an acceptable state."[66] The polluter shall thus assume the expenses of carrying out the measures specified by public authorities to protect the environment. The precise meaning of the "polluter-pays" principle remains open to interpretation, which may be based on different situations.[67] However, it emphasises that the person who pollutes the environment must pay for the damages and consequential costs he has caused, assuming the polluter in a given case is readily identifiable.

In 1993, the European Commission's Green Paper started to use the "polluter-pays" principle to justify civil liability and asserted that the "polluter pays" principle was also applicable because "civil liability is a means for making parties causing pollution to pay for damage that resulted."[68] In this sense, it was likely

---

[65] Bergkamp, Lucas, *Liability and Environment* (2001), p. 15.

[66] This is the definition given by OECD, cited by Bergkamp, Lucas, *ibid.*, p. 15.

[67] Sands, Philippe J., *Principles of International Environmental Law I: Frameworks, Standards and Implementation* (1995), p. 214. The first international instrument to refer expressly to the "polluter pays" principle is the 1972 OECD (Organisation for Economic Co-operation and Development) Council Recommendation on Guiding Principles Concerning the International Economic Aspects of Environmental Policies, which endorsed the "Polluter-pays" principle to allocate costs of pollution prevention and control measures, encourage rational use of environmental resources and avoid distortions in international trade and investment. Following OECD resolutions and recommendations, the principle was incorporated in the Single European Act and, in general, European jurisdictions have been paying increasing attention to the principle, although there is less support outside the European States.

[68] Commission of the European Communities. Communication from the Commission to the Council and Parliament and the Economic and Social Committee: Green

that the "polluter pays" principle can be invoked to require the polluter to compensate for the environmental damage resulting from his activities.[69] Some conventions were influenced by this idea and had a strong desire to channel compensation to those responsible for the activity causing damage to the victims[70]: for instance, the Vienna Convention on Civil Liability for Nuclear Damage.[71]

It is relevant to ask whether the strict liability rule in the international civil liability convention can be justified by the "polluter pays" principle. The most obvious difficulty in this regard is the identification of the polluter. Most conventions make the owner of the ship strictly liable for oil-pollution risk. However, in an incident such as an oil spill, pollution damage is frequently caused by the fault of persons other than the owner of the ship from which oil escapes,[72] this was furthermore analysed as follows:

> "...it is a notion which may need to be applied with caution in the more complex context of shipping...Shipping casualties are not infrequently caused by the fault of persons other than the owner of the ship from which oil escapes. By definition, strict liability makes the shipowner responsible for something which is not his fault, and for which he would not otherwise be liable. To refer to him indiscriminately as 'the polluter' is potentially misleading and may encourage an unduly emotive approach to the issue."[73]

As a matter of fact, the interests involved in maritime transport are quite sensitive and tend towards a sharing of costs on the basis of equity or otherwise in most cases. For instance, the Preamble to the 1971 Fund Convention indicates that the economic consequences of oil-pollution damage should be borne by the shipping industry and oil-cargo interests together. Some academics have thus considered the shipping industry and oil-cargo interests as "polluter" in a collective sense. The concrete application of "polluter pays" principle is thus achieved by contributions from the shipping industry and oil-cargo industry. However, this idea is not applicable to the Bunkers Convention since there is a major difference between carriage of oil cargo and non-oil cargo by sea. In the latter case, only the shipowner is required to be liable for pollution damage under the Bunkers Convention. Therefore, this author agrees that although the shipowner is strictly liable under the Bunkers Convention, it may not be appropriate to consider and define him as "the polluter".

Paper on Remedying Environmental Damage. Brussels, 14 May 1993, COM (93)47 final. In Bergkamp, Lucas, *supra,* note 65, p. 16, F.N.46.

69 Bergkamp, Lucas, *ibid.*, p. 16.

70 Sands, Philippe J., *supra*, note 67, p. 214.

71 See <http://www.iaea.org/Publications/Infcircs/1996/inf500.shtml>, the convention was adopted on 21 May 1963 and entered into force on 12 November 1977.

72 See IMO LEG 74/4/2.

73 See IMO LEG74/4/2.

II. The significance of Art. 7(8) and financial standing of providers of insurance or financial security

The financial standing of providers of insurance or other financial security is of great concern. It is especially due to the reason that no second-tier compensation fund such as the IOPC Fund exists for victims of bunker pollution. Therefore, it is imperative to ensure that the provider of insurance or financial security has a sound financial standing and thus ensure its capacity to pay compensation once the incident occurs. The term of "financial standing" is not defined in the Bunkers Convention; however, it may follow from Article 7(9) that the insurer or guarantor shall be "financially capable of meeting the obligations imposed by this Convention".[74] In brief, the insurer or the guarantor shall be solvent and maintain his financial capacity to pay claims at the time when the certificate is issued by an appropriate authority, or arguably during the period of validity of the certificate.[75]

Article 7(8) is of particular significance. It is important also to note that no similar provisions exist in the CLCs. The whole paragraph reads:

> "Nothing in this Convention shall be construed as preventing a State Party from relying on information obtained from other States or the Organisation or other international organisations relating to the financial standing of providers of insurance or financial security for the purposes of this Convention. In such cases, the State Party relying on such information is not relieved of its responsibility as a State issuing the certificate required by paragraph 2."[76]

As seen, it is important for a State Party to have the relevant information regarding the financial standing of providers of insurance or financial security. However, the Convention does not in itself impose any obligation on any State, whether it is a State Party to the Convention or not, or related organisations, to assist a State Party in obtaining such kinds of information and ensuring its accuracy. The same paragraph furthermore requires that the State Party relying on the information available from other States or relevant organisations is not relieved of its responsibility as a State issuing the certificate required by paragraph 2 of the same Article.[77] Therefore, a State Party, having issued a certificate in accordance with the information from the other State, may still have to be responsible if it turns out that the insurer in question is not financially capable in the end.

By virtue of Article 7(7), the State of the ship's registry shall determine "the conditions of issue".[78] The phrase of "conditions of issue" may be understood extensively to include the reasonable financial standing of providers of insurance or financial security. However, it does not expressly indicate that there is a duty on

---

[74] The Bunkers Convention, Art.7(9).

[75] The insurer must have sound financial standing at the time of issuing the certificate, and this author also believes that it is necessary for the insurer to be able financially to meet the economic consequences of pollution during the period of validity of the certificate.

[76] The Bunkers Convention, Art.7(8).

[77] The Bunkers Convention, Art.7(2).

[78] The Bunkers Convention, Art.7(7).

the State of the ship's registry to evaluate the financial soundness of providers. If this paragraph is read in combination with Article 7(2), the name and address of the insurer of guarantor but not the financial standing of the insurance company shall be contained in the certificate when issuing the certificate. It is, nevertheless, not clear whether and how the State of the ship's registry or any State Party, if a ship is not registered in a State Party, shall evaluate the financial standing of the insurer or the guarantor.

Article 7(9) furthermore provides that, if a State Party believes that the insurer or guarantor named in the insurance certificate is not financially capable of meeting the obligation of this Convention, it may at any time request consultation with the issuing or certifying States.[79] This paragraph uses a weak term – "may". Therefore, there is no obligation on a State Party to evaluate the financial standing of the insurer, even if it believes that the insurer or guarantor named in the certificate is not financially capable.

From the above analysis of the relevant paragraphs, it is clear that many aspects regarding the financial standing of providers of insurance or financial security are not mentioned or are uncertain under the Bunkers Convention. This is due to the following reasons: first, the Bunkers Convention shall not specify the source or types of insurance, since, for instance, this may come into conflict with WTO regulations and European competition law.[80] Secondly, it would be inconsistent with practice if the Bunkers Convention virtually laid down the standards of financial standing of the insurer. The reason derives from the nature of insurance practice. The insurance company is routinely supervised by the regulatory authority of its principal place of business. In addition, its financial standing is regulated by national law. For instance, in Germany, the supervision of insurance companies is regulated by federal law and state ("Länder") law.[81]

It may happen, in practice, that substandard insurers issue worthless insurance policies to substandard shipowners.[82] It has also happened that some insurance providers have collapsed during the insurance period. It might be irrelevant if the financial standing of insurance providers is poor for their business. However, it would be important if the poor financial standing of any insurance providers influenced the insurance arrangement for the purpose of the Bunkers Convention, i.e. the insurance company is financially incapable of paying any compensation once a bunker-oil spill occurs. This may rarely happen, but, once the Bunkers Convention is in force, if and when this happens, the certificate, which was issued according to the former conditions of issue, should be withdrawn by the State that had issued the certificate. The certificate should not be presented any more and an

[79] The Bunkers Convention, Art.7(9).

[80] Gyselen, Luc, 'P&I Insurance: the European Commission's Decision Concerning the Agreements of the International Group of P&I Clubs', In: Huybrechts, Marc/ Hooydonk, Eric Van/Dieryck, Christian (eds.), *Marine Insurance at the Turn of the Millennium* (1999), Volume 1, pp. 181-202.

[81] Gabriel, Moss (ed.), *Cross-frontier Insolvency of Insurance Companies* (2001), p. 249.

[82] Bennett, Paul, *supra*, note 56, at 19.

alternative arrangement should be made. The Bunkers Convention does not contain any provision regarding the consequences if the certificate were still presented in this case. It is therefore up to each State Party to solve the issue according to specific national law or policies. The ship concerned might have to be detained according to national law. The States may take the matter up and negotiate with other States concerned through any diplomatic channel.

It is a fact that a vessel can be insured in different countries in today's globalised shipping industry. Most of the ships are insured by P&I Clubs with a good reputation; however, some of them may obtain insurance from a P&I Club outside the International Group of P&I Clubs or from a commercial insurer.[83] Therefore, it would be more secure if the financial standing of the providers were expressly required under the Bunkers Convention. When insurance or other financial security is presented with a request to the appropriate authority of a State Party to issue the insurance certificate, it might be necessary not only to ascertain the monetary amount of security as required under the Bunkers Convention,[84] but also to evaluate and anticipate the possible financial standing of the providers at the time or even during the period of validity of the certificate. However, the provisions in the Bunkers Convention concerning the latter are weak. Nevertheless, the name and principal place of business of insurer or other person giving security and, where appropriate, place of business where the insurance or security is established are required to be entered in the certificate in any case.[85] It is most likely that a State Party will rely on any information from other States where the providers have their principal place of business before issuing the certificate, but it is not certain to what extent a State Party has access to such kind of information. This will be more uncertain if the insurance company is not a State Party to the Bunkers Convention.[86]

It might also be necessary to highlight that the providers of insurance or other financial security shall be capable of meeting their financial obligation for the purposes of the Bunkers Convention at the time of the incident. This also applies to P&I Clubs. As a general matter, a P&I Club is solvent as long as it still has solvent members. The Club may, however, also go bankrupt. The *Deifovos* case in Norway partly illustrated this possibility. The said incident took place on 25 January 1981. The costs of the whole pollution damage were US$2,811,000, with

---

83 Gyselen, Luc, *supra*, note 80, p. 182: "The remaining tonnage is either not insured at all or is insured by small independent P&I mutuals or commercial insurers operating in 'niche' segments of the P&I market (e. g. covering relatively low risks such as dry cargo, coastal or fishing vessels). Some Lloyd's syndicates have recently entered the market…"

84 The Bunkers Convention, Art. 7(2).

85 The Bunkers Convention, Art. 7(2).

86 The Bunkers Convention, Art. 7(9): "…A state Party may at any time request consultation with the issuing or certifying State should it believe that the insurer or guarantor named in the insurance certificate is not financially capable of meeting the obligations imposed by this Convention." However, it may happen that the insurer or guarantor named in the insurance certificate does not belong to the issuing or certifying State.

a limitation amount of US$1,520,000. The owners and the P&I Club were at the time believed to be of poor financial standing, and the claimant was advised at the time that the P&I Club had not properly reinsured. This claim was financially abandoned.[87]

III. Mutuality

Mutuality is an outstanding feature of P&I Clubs. It explains why the members commit themselves to furnishing funds to pay claims for which they or their fellow members are liable. More discussion will focus on: (1) the meaning of mutuality and (2) the role of mutuality.

1. The meaning of mutuality

"Mutuality" is usually defined as a reciprocal relationship between interdependent entities. "Mutuality" in the P&I Club denotes that firstly, all current members of the Club are the assureds and insurers at the same time, although it appears that the prospective member enters into a contract of insurance with the corporate insurer – the P&I Club.[88] Secondly, all current members share the interests as well as the risks with one another in the Club. As far as the liability for oil-pollution risk is concerned, mutuality means that each assured, the shipowner in this case, is liable to meet a share of the costs of pollution damage caused by himself and all other members of the Club.[89]

As a matter of fact, the relationship between the assured and the insurer within the P&I Club is different from that under other marine insurance arrangements. In other types of marine insurance, the internal relationship between the assured and the insurer is primarily contractual in nature. There are mutual and reciprocal obligations, rights and powers between the assured and the underwriter as defined in the terms of the insurance contract.[90] The underwriter simply undertakes to indemnify the assured, in such a manner and to the extent as agreed under the contract of insurance. Moreover, even if different assured parties are insured under one underwriter in the same insurance company, or purchase a similar type of insurance, there is no mutual interest among those assured parties.

In addition, mutuality not only exists among the members of a Club, but also among P&I Clubs. The Clubs in the International Group of P&I Clubs operate under a claim-sharing arrangement. They share claims in excess of a certain amount with one another under such a special arrangement.

---

87 See IMO LEG 75/5/1.

88 Hill, Christopher/Robertson, Bill/Hazelwood, Steven J., *Practical Guides: Introduction to P&I* (1996), p. 11.

89 Bennett, Paul, *supra*, note 56, at 14.

90 Thomas, David Rhidian (ed.), *The Modern Law of Marine Insurance* (2002), p. 1.

2. *Role of mutuality*

"Mutuality" embodies at least two merits: first, it enables the Club to absorb large claims; secondly, it provides an incentive for the members of the Club to reduce risk.

a) The ability to absorb large claims

The system of levying calls rather than charging premiums is central to the concept of mutuality.[91] Therefore, in theory, a Club has the potential ability to absorb large claims since the members in the Club have the obligation to pay a supplementary call to the extent that the advance call is insufficient to cover claims.

As mentioned, the "call" system is comprised of the advance call, the supplementary call and the overspill call in "catastrophic" years. The system is regulated by the Club Rules.[92] In practice, although advance calls are announced at the beginning of each policy year, they can be paid by instalments during the policy year.[93] The committee of directors of the Club has the power to decide to levy supplementary or additional calls. They also decide upon the percentage to be applied to the supplementary call.[94] The committee or the managers generally indicate an estimate of the percentage at which it is hoped that any supplementary call or calls will be levied. However, it is made clear by the Club Rules that any such estimation shall be without prejudice to the right of the committee or managers to levy supplementary calls at a higher or lower percentage than what was indicated and the Club disclaims any liability whatsoever arising as a result of any estimate or in respect of any "error, omission or inaccuracy" contained in an estimate.[95] It may happen, for example, when the estimate of the managers proves to be too conservative. The Club Rules also prescribe how calls should be paid and provide the managers and committee with a wide discretion as to the number of instalments and on what dates and in which currency they should be paid.[96] Therefore, such a relatively flexible call system reflects a mutual characteristic of P&I Clubs.

Additionally, as indicated, mutuality also exists among the Clubs. The structure known as the "International Group of P&I Clubs" arranges a "claims-sharing agreement",[97] which ensures that commercial reinsurance is viable and large

---

91 Hazelwood, Steven J., *supra*, note 36, p. 121, footnote omitted; see also Chapter 2, Section C.I.4.c).

92 See Chapter 2, Section C.I.4.c).

93 Hazelwood, Steven J., *supra*, note 36, p. 122.

94 Hazelwood, Steven J., *ibid.*

95 Hazelwood, Steven J., *ibid*, p. 123.

96 *Ibid.*

97 See Chapter 2, Section C.I.4.d). Some information is stated by the UK P&I Club at: <http://www.ukpandi.com/ukpandi/infopool.nsf/HTML/About_IG> (visited 30 June 2005): "Although the Clubs compete with each other for business, they have found it beneficial to pool their larger risks under the auspices of the International Group. This pooling is regulated by a contractual agreement which defines the

claims can be met. This agreement thus displays an advantage of P&I insurance over other insurance. If claims made on one Club are in excess of a certain amount, the extra will be shared proportionately among all Clubs in the group pool. However, this relationship does not exist in hull insurance, which usually contains a limitation in the insurance contract with regard to the amount insured.

b) Possible motivation for risk minimisation

To some extent, any type of insurance may be able to minimise or reduce risks "by requiring insureds to adopt appropriate safety measures and by monitoring implementation efforts".[98] In particular, "mutuality" within P&I insurance means that each shipowner is liable to meet a share of the losses caused by himself and all other members of the Club; in other words, it has the effect of internalising costs and expenses on the individual member. In combination with the non-profit characteristic of P&I insurance, the interests of the assured, i.e. shipowners, are more closely related. This may provide an incentive to the shipowner to minimise risk.

Once a Club has to indemnify claims on a large scale in a policy year, it is the individual member who will receive a high "call". The Donaldson Report in the *Braer* oil-spill case mentioned that each P&I Club member had "an interest in seeing that the claims of his fellow members are as infrequent and as small as possible".[99] Because of this, the report went on to argue that the P&I Clubs are one of the best-placed private actors for encouraging higher safety and environmental performance in the shipping industry.[100] One may argue that mutuality may also have an opposite effect. Like any other types of insurance, P&I insurance might lead the individual shipowner to take less care, since he knows that the costs of any incident will be shared amongst all members in the Club once insured. However, the shipowner has also to be aware that, although the high cost claims in his incident will be absorbed by other members of the Club, over the longer term every member in the Club will have to pay higher calls.

There are institutional arrangements in the Clubs. The Club sets rates that are commensurate with the risks of each member. Furthermore, once a member is admitted to a Club, a database will be set up to record the claims, the causes of those claims and the shipowner's past and present attitude to the issues of safety.[101] However, it may also occur that a substandard ship jumps from one Club to another. Therefore, not only shall each shipowner in the Club be concerned about the performance of other members, but each Club in the International Group

risks that are to be pooled and exactly how these are to be shared between the participating Clubs..."

98 Richardson, Benjamin J., 'Mandating Environmental Liability Insurance', 12 *Duke Envtl. L. & Pol'y F.* 293, at 296.

99 Bennett, Paul, *supra*, note 56, at 14, cited from Donaldson J., 'Safer ships, cleaner seas: report of Lord Donaldson's inquiry into the prevention of pollution from merchant shipping', London: HMSO.1994.

100 Bennett, Paul, *ibid.*

101 *Ibid.*, p. 17.

shall also be concerned about the performance of other Clubs.[102] In this respect, the International Group has taken relevant measures to limit unfair competition between Clubs. The agreement called "International Group Agreement" is claimed to prevent "poorly performing members from escaping the financial penalties of their actions."[103] It protects a Club's existing membership against unreasonably low rates being offered by other Clubs.[104]

E. Liability insurance and compensation fund

The traditional method of compensating for damage is through liability insurance, but the inherent limitation of insurance has prompted the development of supplementary sources.[105] The compensation may be from: (1) the liable person or his insurer, such as a P&I Club; or (2) a fund scheme or similar instrument. Currently, the preferred approach under international civil liability conventions appears to be the combination of liability insurance with a compensation fund. The two-tier compensation system to remedy tanker-oil pollution damage at sea is often cited as an example.

It is normal that those who caused pollution damage shall be liable for it. However, it has to be admitted that the undertaking of the liability has become more costly, so that it is rare for the person actually at fault to be the person who pays. Third-party liability insurance is thus commonly utilised to cover pollution damage. The use of the insurance mechanism requires that the risk is insurable and the insurance is available. The idea of compulsory insurance introduced into international civil liability conventions provides victims with a more secure means of compensation.

A compensation fund is set up to compensate for pollution damage if needed. The establishment of a fund is considered necessary in at least two circumstances. The first is when the person liable for pollution damage is unknown or is exempted from liability under the established liability regime;[106] the fund scheme should accordingly substitute the liable person by paying compensation to any person suffering pollution damage. Secondly, if the person liable is short of financial resources and incapable of meeting his obligation, the fund scheme should play a supplementary role in such a circumstance. The typical method is to create a fund with financial contributions from relevant industries. The contributing industries are usually those which are engaged in the activities likely to produce pollution damage of the kind envisaged by such a liability regime.

---

102 *Ibid.*, p. 17.

103 *Ibid.*, p. 18.

104 Gold, Edgar, *Gard Handbook on P&I Insurance* (2002), p. 111. See also *ibid.*, p. 17: "...if a shipowner switches Clubs, the new Club must not undercut the old Club for at least 1 year."

105 Larsson, Maria-Louise, *The Law of Environmental Damage: Liability and Reparation* (1997), p. 561.

106 It is clear that there are exceptions even to strict liability.

The compensation fund may be of various kinds and operate in different ways to ensure its efficacy as a remedy. In general, the fund has certain features: first, the contribution rules can be designed to reduce and distribute (in time and over a number of persons) any risks that result from the covered activities. Secondly, a fund can require some upfront payments and will thus reduce the risk of unavailability or insolvency. For example, some one-ship companies have become insolvent when ordered by a court to pay their liability.[107] Thirdly, once a fund scheme is established, it can make compensation available on the basis of need. For instance, the fund could provide timely and adequate resources for carrying out preventive measures,[108] where the liability insurance may usually not do so. Fourthly, the administrative cost associated with a liability insurance system may be much greater than that of running a fund. Fifthly, liability rules and doctrines will be difficult to apply to some unforeseeable or multi-causal damage, whereas a fund can be governed by a specific set of rules. Finally, from a moral perspective, compensation schemes may be preferable to liability regimes.[109] However, the plans and arrangement of a compensation fund will involve massive bureaucratic participation of the States.

However, the suggestion of a compensation fund or a similar arrangement along with the Bunkers Convention in this research thesis is not meant to replace liability insurance. Any compensation fund should contribute to ensuring or strengthening the availability and adequacy of compensation. It would be practical for a compensation fund, once it could be set up, to be only either supplementary or complementary in nature or both for any claims under the Bunkers Convention.

F. Adequacy and other types of compensation

The purpose of adopting the Bunkers Convention is to preclude situations where compensation might be unavailable or not adequately available from the liable person or his liability insurer. Therefore, a situation in which pollution damage remains totally or partially uncompensated for will be contrary to the objective of the Bunkers Convention. It has chosen the shipowner and his liability insurer as the sole source of compensation. This leads to the following examination of the effectiveness of the current insurance arrangement as a means of guaranteeing compensation and the discussion of other possible sources of compensation.

It is clear that the risk of bunker-oil spill may be very high. The spillage of bunker oil can have devastating consequences for the marine environment and its dependent industries.[110] It is possible that claims arising out of a bunker-oil spill will exceed the limitation provided in the current Bunkers Convention. Since there

---

[107] Bergkamp, Lucas, *supra*, note 65, p. 229. The author discusses the fund scheme in general.

[108] As illustrated in Chapter 2, the IOPC Fund provides funds for undertaking preventive measures.

[109] For more details in Bergkamp, Lucas, *supra*, note 65, pp. 225-230.

[110] See Chapter 1, Section B.III.

is a lack of data to be used as the basis for estimating total liabilities and the minimum amount of compensation needed for a bunker-oil spill incident, it is, thus, difficult to evaluate whether the present system in the Bunkers Convention satisfies the need for adequate compensation.

The availability of other compensation sources in some circumstances may be necessary. The reasons for considering other compensation sources were three-fold: first, to remedy the potential insolvency of the shipowner; secondly, to provide supplementary compensation above the amount contained in the current limitation regime under the Bunkers Convention; thirdly, to spread the loss and to balance the industries' interests. More specifically, the first two objectives would correspond to the purpose of the Bunkers Convention to ensure the payment of adequate, prompt and effective compensation. Besides, they could also spread the risks among the main parties involved in maritime transportation; in other words, it could balance the interests involved. As pointed out by one author, "...the history of oil pollution legislation, as with any other legal system, is the history of a quest for balance in weighing the conflicting interests of the parties involved."[111] Therefore, it would also be necessary to seek a balance of interests under the Bunkers Convention. It is, however, important to choose a proper manner in which the financial burden can be apportioned between the interests. For this purpose, the discussions below will mainly include the parties such as the shipping industry and cargo industry. The possibility of seeking compensation from State Parties to the Bunkers Convention will also be considered.

I. The willingness of P&I Clubs to increase their coverage limit

As the liability insurance taken out by the registered owner will be the only guaranteed source of compensation for pollution victims and the Clubs will be the main liability insurer, it is of relevant interest to ask how willing the Club would be to increase its coverage limit for the oil-pollution risk to its maximum possible extent.

The limitation that applies to liability of the shipowner is laid out in the Bunkers Convention by reference to "any applicable national or international regime".[112] Due to the uncertain nature of this provision, the shipowner may be subject to unlimited liability if the national legislation in a contracting State requires it.[113] The shipowner may thus hope his liability insurer offers as high an insurance coverage as possible for the bunker-oil pollution liability.

The Club's ability to provide insurance cover depends largely on its ability to purchase reinsurance from the commercial insurance market. Most of this is

---

[111] Wu, Chao, *Pollution from the Carriage of Oil by Sea: Liability and Compensation* (1996), p. 3.

[112] The Bunkers Convention, Art.6.

[113] Tsimplis, Michael N., 'The Bunker Pollution Convention 2001: Completing and Harmonizing the Liability Regime for Oil pollution from Ships?' 1 *Lloyd's Mar. & Com. L.Q.* (2005), pp. 83-100, at 83.

purchased from Lloyd's. In effect, the International Group has recently increased reinsurance protection for oil pollution from US$500 million to US$1,000 million.[114] The Clubs are exposed to each accident or occurrence in respect of each ship entered by or on behalf of an owner for US$1,000 million (US$1 billion), but no more than this amount. Oil-pollution risks under P&I insurance have been limited since 1970. The limitation was imposed mainly due to the concern about the possibly disastrous consequence of an oil-spill incident. The limitation of liability established by the Clubs is not necessarily a disadvantage per se; however, if there is only liability insurance available for compensation, the limitation itself will be very disadvantageous to pollution victims.

When selling insurance for tanker-oil pollution liability, a P&I Club has to take into account the fact that the number of tanker owners who may incur catastrophic liabilities only represents a portion of the members in the Club; if the limitation is set too high, it may be difficult to balance the interests between the tanker-owner members and non-tanker-owner members of the Club. Therefore, there is a need for a strict limitation on tanker liability insurance. However, if almost all members of the Club are required to take out insurance for similar bunker-oil spill liability, it would be possible for the Club to think about providing a more favourable limit in this respect.

II. The compensation paid by the cargo interests

It seems to be unfair to impose liability on the cargo interests as the bunker oil pollution is a danger inherent in the carriage and the cargo interests usually have no operational control over the vessel.[115] Nevertheless, the fact that it is suggested that cargo interests are liable for compensation is due to the following: first, pollution victims need to be adequately compensated; secondly, cargo interests represent one of the main beneficiaries of the carriage of goods by sea, and so they should also assume the economic consequence of oil-pollution damage. Thirdly, compensation from cargo interests can relieve the shipowners of any additional financial burden imposed on them.

We might require cargo interests to either take out liability insurance or establish a fund. Since varied cargo interests may be involved in the carriage of goods by sea, it would be impossible to allocate this requirement to all relevant cargo interests. Therefore, one particular party should be selected from the cargo interests and he would be required to maintain insurance or other financial security according to the relevant provision.[116] We might impose the task of identifying this person on the shipowner, since he could be deemed to be the party to have constant contacts with the cargo interests. If the shipowner failed to identify the

---

[114] The information is from: <http://www.ukpandi.com/ukpandi/Infopool.nsf/HTML/E8EBE7157D236C5080256DB30055E4BD?> (visit 23 November 2005), see Chapter 2, Section C.II.1.b).

[115] See Chapter 3, Section D.I.2.

[116] The Bunkers Convention, Art. 7(1).

liable person from the cargo interests, the shipowner would have to stand in the shoes of this liable person whom he failed to identify.[117] To establish a fund would be an alternative means to inviting cargo interests to share the financial burden in bunker-oil spill incidents. The manner of handling a claim under this desired fund may be designed to be similar to what is in the IOPC Fund. However, the question is: who should be the contributors to this desired fund? In effect, in both cases, feasibility would be hindered by the difficulty in identifying the liable person to contribute.

Cargo interests can involve any person who has an interest in the cargo that is shipped. Except the carrier[118], it may include the shipper, the consignee and the owner for the time being of the cargo and also other types of ocean transport intermediary.[119] These terms are used frequently in the contract of carriage by sea. "Shipper" is the party who supplies the cargo and initials a contract of carriage with a carrier.[120] "Consignee" is the party who is entitled to take delivery of the goods. His identity is often noted on the bill of lading.

To consider imposing liability on cargo interests, it might only be plausible to impose liability on a particular type of cargo interests who could be most easily identifiable. For instance, the contributions made to the IOPC Fund are clearly from oil importers.[121] However, since different types of cargo are involved, it would not be feasible to follow the same approach as in the IOPC Fund for bunker oil pollution liability. There are millions of importers or exporters of different

117 This idea follows the Irish proposal in the discussion of the 1969 CLC, see IMCO, O.R., 1969, pp. 446-452.

118 See Hague Rules 1924, reprinted in: *Lloyd's Shipping Law Library*: *The Ratification of Maritime Conventions* (2004), Vol.3, II 5.10, Art. 1(a), "'Carrier' includes the owner of the vessel or the charterer who enters into a contract of carriage with a shipper." Hamburg Rules 1978 also has the definition, reprinted in: *Lloyd's Shipping Law Library*: *The Ratification of Maritime Conventions* (2004), Vol.3, II.5.220, "Carrier" refers to "any person by whom or in whose name a contract of carriage of goods by sea has been concluded with a shipper."

119 A term used in the United States includes both the NVOCC and the ocean freight forwarder. The former denotes the "Non-Vessel-Operating Common Carrier", which means a common carrier that does not operate the vessel by which the ocean transportation is provided and is a shipper in its relationship with an ocean common carrier. The "freight forwarder" is a bit complicated. At times, the freight forwarder acts as a principal contractor in respect of the shipper and bears the responsibilities of a common carrier. At other times, the freight forwarder acts merely as an agent of the shipper, with the obligation to exercise reasonable care and skill. The definitions are from *Tetley's Glossary of Maritime Law*, available at: <http://www.mcgill.ca/maritimelaw/glossaries/maritime/> (visited 24 November 2005).

120 "Shipper" is defined in the Hamburg Rules 1978 as "any person by whom or in whose name or on whose behalf a contract of carriage of goods by sea has been concluded with a carrier, or any person by whom or in whose name or on whose behalf the goods are actually delivered to the carrier in relation to the contract of carriage by sea."

121 The Fund Convention, Art.10(1).

types of cargo every day, so it would be difficult to require either the importer or the exporter to contribute to and set up a compensation fund, because huge bureaucratic machinery would be needed in the Member States. To sum up, the intention to impose a compensation obligation on cargo interests will be hampered by the practical problem of identifying contributors among those cargo interests.

III. Compensation paid by other jointly liable persons

1. "Joint and several liability" rule in relation to the compensation purpose

At the time when the bunkers convention was conceived and debated, some attention was focused on sharing the burden of liability where it could be shared. Therefore, the definition of "shipowner" in the Bunkers Convention is broad, and all persons involved are accordingly liable for pollution damage.[122] This position is fairly intensified by the "joint and several liability" provision, which provides that where more than one person is liable for pollution damage, their liability shall be joint and several.[123]

A "joint and several liability" is a contributory liability rule which entitles the claimant to recover the amount of compensation from any one of multiple defendant persons or from any combination of them. Therefore, even if one is insolvent or exonerated from liability, claimants can collect the amount of compensation from other liable persons including the registered owner, bareboat charterer, manager and operator of the ship.[124]

The "joint and several liability" provision reflects a strong desire of the Bunkers Convention to collect any possible source of compensation for pollution victims. However, compensation from the liable persons other than the registered owner is not guaranteed, since other persons are not required to take out insurance or arrange other financial security to guarantee their liability. Maintaining liability insurance or other financial security is optional for those parties. In addition, the scope and extent of the insurance coverage can also be chosen by them. Therefore, the claimants may not receive compensation from those persons if they are insolvent in the absence of insurance. Even if they take out insurance, it might still be an inadequate guarantee in the case of these persons being underinsured. Therefore, it would be more secure for eligible claimants if all persons involved could provide some form of financial security for compensation.

[122] The Bunkers Convention, Art. 1(3).
[123] The Bunkers Convention, Art. 3(2).
[124] The Bunkers Convention, Art. 1(3).

2. *Bareboat charterer*

a) "Demise charterer" or "bareboat charterer"?

There were discussions on the concepts of "demise charterer" and "bareboat charterer". The combined wording of "bareboat and demise" was initially included in the draft convention in the definition of "shipowner", following the notion that the terms "demise charterer" and "bareboat charterer" should be different.[125] However, most delegations expressed a preference for only maintaining a reference to "bareboat charterer" and dropping the term "demise charterer". It was also pointed out that the reference to "bareboat charterer" was not intended to exclude the idea of "demise charterer" but that, in modern terminology, the term "bareboat charterer" could cover the concept of "demise charterer".[126] It is not uncommon for these two phrases "bareboat charter" and "demise charter" to be used interchangeably.

The reason for holding the bareboat charterer liable is based on the characteristic of the bareboat charterer. In practice, the bareboat charterer steps into the shoes of the shipowner and the operator and assumes control over the management and operation of the vessel. The bareboat charterer is responsible for the ship. For instance, if the ship is involved in a collision, the charterer must answer to the owner for any damage incurred.[127] It is, therefore, natural that the bareboat charterer shall be exposed to liability for oil pollution. Under general maritime law, the bareboat charterer, as owner *pro hac vice*, is subject to personal liability for pollution damage sustained as a result of the fault or neglect of the vessel's crew.[128]

Where pollution damage exceeds the limitation applicable to the registered owner or where the registered owner fails to respond to the damage, pollution victims may look to other liable persons such as the bareboat charterer for the recovery of any shortfall.[129]

b) The insurance of the bareboat charterer

The owner of the ship usually seeks to require the charterers to arrange and pay for suitable insurance as agreed and specified in the bareboat charter party.[130] In most circumstances, the protection and indemnity risks are therefore required to be insured at the expense of the bareboat charterer. For instance, Clause 13 of the

---

125 See IMO LEG 79/6/1.
126 See IMO LEG 81/WP. 2.
127 Gold, Edgar/Chircop, Aldo/Kindred, Hugh, *Canadian Maritime Law* (2003), p. 380.
128 Anderson, Charles B./DelaRue, Colin, M., 'Liability of Charterers and Cargo Owners for Pollution from Ships', 26 *Tul. Mar. L. J.*, pp. 1-60, at 9.
129 The Bunkers Convention, Art. 1(3), Art.3(1).
130 Davis, Mark, *Bareboat Charter* (2000), p. 65.

Bimco Barecon 89 Form, which is used for bareboat chartering in practice,[131] reads:

> "Insurance, repairs and classification:
> (b) During the Charter period the Vessel shall be kept insured by the Charterers at their expense against Protection and Indemnity risks in such form as the Owners shall in writing approve which approval shall not be unreasonably withheld. If the Charterers fail to arrange and keep any of the insurance provided for under the provisions of sub-clause (b) in the manner described therein, the Owners shall notify the Charterers whereupon the Charterers shall rectify the position within seven running days, failing which the Owners shall have the right to withdraw the Vessel from the service of the Charterers without prejudice to any claim the Owners may otherwise have against the Charterers."

In practice, although the bareboat charterer is not required to take out insurance under the Bunkers Convention, he will possibly be required to take out insurance under the charter party. Moreover, it is common for the liabilities of owners and charterers of demise-chartered ships to be jointly insured under the same P&I cover in practice.[132] In other words, both charterers and owners may join a P&I Club under a "family arrangement", whereby both shipowners and charterers enter the same Club as co-assured parties and co-members.[133]

It is true that the owner of the demise-chartered vessel remains in need of liability insurance, particularly in respect of those claims for which he is still liable as a shipowner. However, in most cases of demise or bareboat chartering, protection and indemnity insurance is to be taken out by the bareboat charterer. Therefore, it would be much more practical if the Bunkers Convention could provide that:

> "The registered owner or the bareboat charterer, when chartering the vessel on bareboat terms, of a ship having a gross tonnage greater than 1,000 registered in a State Party shall be required to maintain insurance or other financial security, such as the guarantee of a bank or similar financial institution, to cover the liability of the registered owner or bareboat charterer for pollution damage in an amount equal to the limits of liability under the applicable national or international limitation regime…"[134]

Under this provision, the obligation of the bareboat charterer to take out insurance would be clear for the purpose of bunker-oil pollution liability.

---

131 In 1989, the Documentary Committee of BIMCO (the Baltic and International Maritime Council) amalgamated the two forms in producing the Barecon 89 form, for use for bareboat chartering. Since its introduction, the Barecon 89 form has replaced its predecessors in popularity and is used for the majority of operating charters. Barecon 89 is reprinted in: Davis, Mark, *ibid.*, pp. 191-203.

132 Anderson, Charles B./DelaRue, Colin, M., *supra*, note 128, at 9, F.N. 40.

133 Hazelwood, Steven J., *supra*, note 36, p. 99.

134 This is based on the Bunkers Convention, Art. 7(1).

IV. Operator and manager

The definition of "operator" is not given in the Bunkers Convention. The term "operator" does not belong to the field of maritime law.[135] In relation to oil pollution, this term first appeared in the discussion of the 1969 CLC. Although liability imposed on the operator was not adopted in the CLCs, reference to the discussions of this term under the 1969 CLC is still worthwhile. The operator was defined in the draft article of the 1969 CLC to mean the person who uses the ship in his own name and mans, equips and supplies it.[136] In any event, the scope of "operator" is subject to judicial determination on a case-by-case basis.[137] The owner of the ship might be the operator himself. However, this is not always the case. It was, however, additionally pointed out that the owner of a ship shall be presumed to be its operator and shall be liable as such unless he can prove that some other person was the operator.[138]

The manager of the ship is also required to take on liability as a consequence of becoming one party in the "shipowner" definition of the Bunkers Convention. The ship manager is regarded as the agent of the owner.[139] He usually takes over from the owner all the main tasks that an owner undertakes, except the commercial operation of the ship, and is responsible in practice for the seaworthiness of the ship, the competence of the crew and the safety of the ship's operations.[140] He is involved in various types of management services which cover all aspects of daily vessel operation comprising technical management. Additionally, due to similar interests and purposes, the manager is often named as the co-assured with the owner in the owner's hull and P&I insurance policies. In these cases, the manager enjoys the benefit of being co-assured, since the insurer for a claim may not sue him in subrogation. However, the manager's negligence can also lead to the loss of the owner's insurance.[141]

Apparently, operators or managers are those who are directly connected with the operation or management of the vessel. The operator and manager shall be the persons liable for pollution, which is in order to avoid the injustice of asking the registered owner to be liable in the case where he is not at fault. Meanwhile, this can also increase claimants' opportunities for recovery. However, it is wise that the Bunkers Convention does not impose compulsory insurance or other financial security obligation on them. The reason has already been given in the discussion of the 1969 CLC, since "placing liability on the operator might place greater administrative burdens on the State issuing a certificate in case of voyage charterers and time charterers (on a short-term basis), since certificates would

[135] O.R.1969 CLC, p. 445.
[136] O.R. 1969 CLC, p. 443.
[137] Chen, Xia, *Limitation of Liability for Maritime Claims* (2001), p. 8.
[138] *Ibid.*
[139] Willingale, Malcolm, *Ship Management* (1998), p. 136.
[140] Willingale, Malcolm, *ibid.*, p. 131.
[141] More read Willingale, Malcolm, *ibid.*

have to be reissued on a frequent basis."[142] The same reason applies to the manager.

The operator and manager can, nevertheless, voluntarily take out insurance to protect their financial interests. In practice, for instance, the International Transport Intermediaries Club (ITIC), which provides professional indemnity insurance for ship agents and shipbrokers also covers risks for claims by third parties against the manager, when the owner's indemnity is inoperative or the owners have gone into liquidation.[143]

V. Liability as the time or voyage charterer

1. Definitions

Charterers of any type were considered as the liable persons during the discussion of a bunkers convention.[144] However, voyage charterers or time charterers are not chosen as the liable persons under the Bunkers Convention

In contrast to the charter party by demise, voyage and time charters are simply different approaches to contracting for the shipowner's provision of the services of the ship.[145] Generally speaking, a voyage charter party is a contract whereby the shipowner agrees to accept the cargo for a voyage or more designated voyages for a consideration called "freight" between two named ports. Comparatively, a time charter party is defined "not by a geographic voyage but by a period of time".[146] Under a time charter party, the shipowner contracts to hire out the vessel at the disposal of the charterer for a period of time as agreed for a consideration called "hire". Apparently, under voyage and time charter parties, although the charterers have a right against the owner to have their goods carried on the vessel, the ownership and possession of the ship remain with the owners through the master and crew, who remain as their servants.[147]

2. Compensation paid by the time or voyage charterer

Article 3(6) provides that nothing in the Convention shall prejudice the owner's right of recourse against the party at fault. This right exists, however, independ-

142 O.R.1969 CLC, p. 457.

143 Willingale, Malcolm, *supra*, note 139, p. 124, p. 131.

144 See IMO LEG 78/WP.3. The discussion of the options for the definition of the concept of shipowner shows that the charterer was chosen as the liable person without further qualification.

145 Gold, Edgar/Chircop, Aldo/Kindred, Hugh, *supra*, note 127, p. 379.

146 Baughen, Simon, *Shipping Law* (2001), p. 172.

147 Davis, Mark, *supra*, note 130, p. 2. However there is a trend that the court is asked to look into the circumstance where the oil spill actually happened. The charterer might in some particular case interfere with the operation of the ship or the charterer had actually exercised operational control over the vessel.

ently of the Bunkers Convention.[148] The phrase "the party" can include the time or voyage charterers. Therefore, in an incident such as a bunker-oil spill, the shipowners shall have the primary responsibility for payment of compensation; it does not, however, prevent him from claiming against the charterer afterwards. It often happens that after a spill the owners investigate if there was a casual link between the spill and breach of the charter party. In practice, one of the greatest risks to the charterers of incurring liability to indemnify the shipowner for pollution has arisen under the safe port or berth warranty contained in most standard forms of charter party.[149]

At the same time, it is also important to know that nothing in the Convention is construed as preventing victims from claiming against the charterers in tort on a national level.

VI. State liability and contributions

"State responsibility" and "State liability" are used simultaneously and sometimes they overlap in the context of international environmental law. However, the latter term may embody more aspects than the former. According to one author, "liability" describes in essence a duty to pay compensation for damage under the concept of *restitutio in integrum*. "Responsibility" encompasses this liability with extended remedies such as injunctions, apologies and so forth, as well as both the obligation not to cause damage and in some cases to take means to prevent it.[150]

The United Nations Convention on the Law of the Sea 1982 lays down, *inter alia*, a framework to regulate all aspects of the protection and conservation of the marine environment. It provides in Article 192 that: "States have the obligation to protect and preserve the marine environment."[151] Moreover, States shall cooperate in the further development of international law relating to responsibility and liability for the assessment of and compensation for damage as well as, where appropriate, development of criteria and procedures for payment of adequate compensation.[152]

In international practice, States have accepted the responsibility for environmental harm. They shall answer for environmental harm caused by activities they have carried out or allowed within their own territory or by activities that are under their control.[153] In this situation, the responsibility or liability of a State cannot arise unless the actor is either an organ or the representative of the State, or a private individual over whom the State possesses legal authority. States may be

---

148 The Bunkers Convention, Art. 3(6).

149 Anderson,Charles B./DelaRue, Colin, M., *supra*, note 128, p. 29.

150 Larsson, Marie-Louise, *The Law of Environmental Damage: Liability and Reparation* (1997), p. 157, any footnote omitted.

151 The Law of the Sea Convention, 1982: Art. 192, reprinted in: *Lloyd's Shipping Law Library*: *The Ratification of Maritime Conventions* (2004), Vol.2, II 170.

152 The Law of the Sea Convention, 1982, Art.235 (3).

153 Francioni, Franesco/Scovazzi, Tullio (eds.), *International Responsibility for Environmental Harm* (1991), p. 15.

held responsible for a vessel's conduct injurious to the marine environment, but only to the extent that the requisite juridical relationship exists between the State and an individual vessel-user.[154]

In general, two types of State liability have been utilised. One makes a State liable when the private actors have failed to fulfil their duty to provide compensation for environmental damage. The convention in the field of nuclear energy is a typical example of this kind of State liability. The private actor is liable up to a certain amount, beyond which the State where the nuclear facility is located pays for the damage, based on a designated distribution formula.[155] The other kind of State liability makes the State liable when it can be proven that it failed in its duty to oversee the operator and that this inadequate supervision resulted in damage. It principally makes the State directly liable.[156]

The question is whether it would be feasible to establish a system of State-funded compensation along with the Bunkers Convention. The impact of this system is primarily to ensure the availability of adequate compensation for pollution victims. Presumably, the Contracting States to the Bunkers Convention would share the burden. It might, however, be more reasonable to impose liability on the flag States in respect of pollution damage caused by their vessels. This could operate as a back-up system of compensation in case the liable person was unable to pay the required compensation.

However, State Parties are not required to contribute to compensation under other international civil liability conventions. In a few cases, flag States have been requested to pay compensation for pollution from oil tankers, but this cannot be treated as supporting the imposition of strict liability on State Parties.[157] It has been observed that the earlier preference of civil liability to State responsibility is due to the fact that the latter has been ineffective as a means of redressing

---

154 The Law of the Sea Convention 1982, Art.235(2): "States shall ensure that recourse is available in accordance with their legal systems for prompt and adequate compensation or other relief in respect of damage caused by pollution of the marine environment by natural or juridical persons under their jurisdiction."

155 For example, *The Convention of 31st January 1963 Supplementary to the Paris Convention of 29th July 1960 as amended by the additional Protocol of 28th January 1964 and by the Protocol of 16th November 1982*. This convention supplemented the measures provided in "*The Convention on third Party Liability in the Field of Nuclear Energy*" with a view to increasing the amount of compensation for damage which might result from the use of nuclear energy for peaceful purposes. The conventions are available at: <http://www.nea.fr/html/law/legal-documents.html> (visited 28 November 2005).

156 The example is from the Eighth Offering for an Annex on Environmental Liability in Antarctica, but it is slightly different. The State would be liable for environmental damage caused by the operator that is not justifiable under the preliminary environmental impact assessment of the project conducted by the State. For more details see Morrison, Fred L./Wolfrum, Rüdiger (eds.), *International, Regional and National Environmental Law* (2000), p. 830.

157 Birnie, Patricia W./Boyle, Alan E., *International Law and the Environment* (1992), p. 291.

international environmental harm.[158] Therefore, it would be not easy to set up such a compensation mechanism.

If flag States were required to share the financial burden,[159] it would still be necessary to emphasise that States' obligation to pay compensation should not prejudice their rights of obtaining reimbursement from the shipowner under its domestic law. Additionally, one needs to distinguish between the obligation to set up a back-up compensation scheme and some basic obligations of States under the Bunkers Convention in such a case. The allocation of liability to States under this back-up arrangement should not affect their liabilities to comply with the obligations under the Bunkers Convention, in particular the obligation to establish and implement a civil liability mechanism, such as establishing the authorities to issue the insurance certificates.[160]

G. Concluding remarks

Once a civil liability convention embodies the issue of compensation, several basic questions must be confronted, such as: first, to whom will compensation be paid if pollution damage occurs? Secondly, how can the compensation source be arranged to ensure that necessary funds will be available? Thirdly, what form can compensation take? Finally, by which criteria is the actual level of compensation determined? Besides, a well-conceived compensation system may need to provide a mechanism for sharing the costs among all those involved in, and benefiting from, the maritime adventure. If the incident in question was unavoidable, there is always a very strong sense for sharing the costs of damage rather than imposing all the losses on the shipowner, particularly in cases where there were no real fault on his part.[161]

In order to answer the above-listed questions, the Bunkers Convention has laid down the relevant provisions. Some of them have been analysed in this chapter. The liability insurance of the registered owner is the main source of compensation under the Bunkers Convention. Its adequacy has been discussed. It is not easy to estimate the extent and scope of pollution damage which may be caused by any

---

158 Morrison, Fred L./Wolfrum, Ruediger (eds.), *supra*, note 156, p. 822, 823.

159 Gauci, Gotthard, *Oil Pollution at Sea* (1997), P.86: "There should be no doubt that this principle of responsibility can be extended to impose liability on flag States in respect of pollution caused by their vessels; however, whether one can go so far as to state that such liability is strict is highly debatable…"

160 Vicuna, Francisco Orrego, 'Responsibility and Liability for Environmental Damage under International Law: Issues and Trends', 10 *Geo. Int'l Envtl. L. Rev.*, pp. 279-308, at 285, 286: "Because environmental regimes usually involve the active cooperation of States in ensuring their effectiveness, the failure of a State to enact appropriate rules and controls to this effect at the domestic level, even if technically not amounting to the breach of an obligation, might engage its international liability if damage ensues as a consequence…"

161 Seward, R. C., 'The Insurance Viewpoint', in: *Limitation of Shipowner's Liability: the New Law* (1986), pp. 161-186, at 163.

disastrous bunker-oil spill incident.[162] Fortunately, to date no disastrous bunker-oil spill incident has been recorded. Any precautionary measures in this respect, including adequate compensation, are nevertheless needed. Compensation from other sources is thus suggested in this chapter. They are intended to compensate eligible pollution victims at least in cases where: first, the liability insurance taken out by the registered owner is not available; and secondly, the shipowner is insolvent or is not identified or where no liability on his side arises. In these cases, the claimants should also be given a right of direct action against the proposed supplementary compensation sources. However, no supplementary compensation source appeared with the advent of the Convention, or at a later stage.

[162] The experience in tanker oil spill incidents may provide an insight. The most recent disaster involving the *Prestige* is a good example. The brief facts of the case were as follows: on November 19, 2002, the 26-year-old, single-hulled oil tanker, *Prestige*, sank and took 50,000 tons of its 77,000 tons of heavy fuel oil down to the bottom of the North Atlantic. It was indicated that the pollution victims of the *Prestige* disaster may never receive full compensation for the losses they have suffered. It can be imagined that if a large amount of fuel oil spills out from a cargo ship such as from a big container ship, the compensation amount envisaged in the Bunkers Convention will be inadequate. Therefore, some precautionary measures need to be considered.

Chapter 5: Limitation of Liability and the Limit of Insurance

A. Introduction

Limitation of liability is a traditional principle in maritime law. After an incident resulting in damage, the owner is to constitute a fund, amounting to the limit of liability if any limitation rule is applicable, and consign it to the Court. From this fund, the claimants will be paid amounts proportionate to their established claims. Once the fund has been set up subsequent to an incident, no person having the claim for damage arising out of that incident shall be entitled to exercise any right against any other assets of the owner in respect of such a claim. A similar approach is utilised in the liability for tanker oil pollution.[1]

Similarly, the Bunkers Convention maintains the shipowner's right to limit his liability. It provides that the liability of the shipowner shall be limited to the amount resulting from any applicable national or international limitation regime.[2] The resolution, which was adopted in this respect with the adoption of the Bunkers Convention, was intended to bring about a basic uniformity of limitation regimes.[3] Therefore, the 1976 LLMC or its 1996 Protocol[4], which has been adopted or transformed into national law in most countries, will most likely be applicable for limiting liability claims from a bunker-oil pollution incident.

As the title of this chapter shows, it will focus on the relationship between the limitation of liability and the limit of insurance. The chapter will first elucidate the limitation of liability rule in the Bunkers Convention. Furthermore, it will discuss

---

1 The 1969 CLC, Art. VI: "1. Where the owner, after an incident, has constituted a fund in accordance with Article V, and is entitled to limit his liability, (a) no person having a claim for pollution damage arising out of that incident shall be entitled to exercise any right against any other assets of the owner in respect of such claim; (b)..."

2 The Bunkers Convention, Art. 6: "Nothing in this Convention shall affect the right of the shipowner and the person or persons providing insurance or other financial security to limit liability under any applicable national or international regime, such as the Convention on Limitation of Liability for Maritime Claims, 1976, as amended."

3 See Chapter 1, Section C.IV.5. The resolution in this respect was adopted to urge all States that had not done so, to ratify or accede to the Protocol of 1996 to amend the Convention on Limitation of Liability for Maritime Claims, 1976.

4 See Chapter 1, Section C.IV.5.

whether the insurance capacity should be considered as a decisive factor in ascertaining the limit of liability.

B. The global limitation of liability system in relation to ships

It is not easy to maintain a set of uniform rules on the limitation of liability on an international level. Each country may have its own legal regime that governs maritime liability limitations. The inherently international nature of shipping, nevertheless, calls for the uniformity of rules on the limitation of liability, which can be realised through international conventions. However, this uniformity depends on many factors. As one author indicates: "such uniformity depends first on the Convention itself, secondly on the manner of its implementation into national laws and on its national effect as against other national laws and thirdly on the consistency with any other Convention which may touch on identical topics."[5]

It was also realised that: "...A major factor in the limitation of liability is the need to recognise and be sensitive to the nature and extent of the risks involved in modern specialist areas of shipping."[6] Accordingly, there are conventions on the limitation of liability for general maritime claims and specific conventions dealing with questions of the limitation of liability in specialised areas or aspects of shipping.

I. Limitation of liability rule in general

The first international convention relating to the limitation of liability for maritime claims was adopted at Brussels in 1924.[7] Prior to this Convention, the shipowner could limit his liability to the value of ship, freight and accessories in many countries.[8] The 1924 Limitation Convention similarly adopted this approach, which entitles the shipowner to limit his liability according to the actual value of

---

5 Jackson, David, "The 1976 Convention and International Uniformity of Rules", in: *The new law: Limitation of Shipowners' Liability* (1986), pp. 126-143, at 126.

6 Beddard, Ralph, 'The Implementation of the Convention', in: *ibid.*, pp. 152-160, at 152.

7 The 1924 Convention is reprinted in: *Lloyd's Shipping Law Library, The Ratification of Maritime Convention* (2004), Vol.2, II. 2.300.

8 Selvig, Erling, "An Introduction to the 1976 Convention", in: *supra,* note 5, pp. 3-17, at 3: "Historically, limits based on the value of the ship long prevailed ..."

the vessel.[9] However, it was a failure. This was mainly due to the fact that it was not adopted by the major shipping nations and thus had little practical value.[10]

The 1924 Limitation Convention was reviewed and was abrogated by the 1957 Convention Relating to the Limitation of the Liability of Owners of Seagoing Ships (the 1957 Convention).[11] The 1957 Convention did not take effect until 1968. This Convention classifies damage in two categories, which are damage to property and personal injury.[12] A fund is required to be constituted according to the 1957 Convention.[13] With respect to the extent of limits, the 1957 Convention offers preferential treatment to claims for personal injury and death. By virtue of Article 3(1) of the 1957 Convention, the amount to which the shipowner may limit his liability differs according to the distinct occasion in question. Where the occurrence has only given rise to property claims, the limitation fund is set at 1,000 francs per ton. In contrast, where only personal claims are involved, the shipowner may limit his liability to an aggregate amount of 3,100 francs per ton. Furthermore, in a case where both personal claims and property claims are involved, the total fund is set at 3,100 francs for each ton of the ship's tonnage, which are divided into two portions. The first portion, amounting to 2,100 francs for each ton of the ship's tonnage, shall be exclusively appropriated to the payment of personal claims. If the claims of personal injury are not fully compensated by the said first portion, the claims in this respect can be in competition with other claims resulting from damage to property and both types of claims share a second portion amounting to 1,000 francs for each ton of the ship's tonnage.[14] The 1957 Convention was criticised in many aspects. One aspect was that the limitation figure could not accommodate the problems of inflation.[15]

---

9 The 1924 Limitation Convention, Art.1: "The liability of the owner of a sea-going vessel is limited to an amount equal to the value of the vessel, the freight, and the accessories of the vessel..."

10 Wu, Chao, *Pollution from the Carriage of Oil by Sea: Liability and Compensation* (1996), p. 33: "This Convention has had little effect since its entry into force in 1931, mainly because it was not adopted by the major shipping nations and therefore had little practical value..." footnotes omitted. Also, Grime, R.P., 'Implementation of the 1976 Limitation Convention: Liability for Maritime Claims', *Marine Policy*, July 1988, pp. 306-313, at 309: "Judged from an international standpoint, the 1924 Convention was not a success." Although the 1924 Convention was ratified or acceded to by 15 States, it did not achieve its objective, most notably because the United Kingdom did not accede to the Convention. See also Selvig, Erling, *supra*, note 8, at 5.

11 The 1957 Convention reprinted in: *Lloyd's Shipping Law Library: The Ratification of Maritime Conventions* (2004), Vol.2, II.2.310.

12 The 1957 Convention, Art.3.1.

13 The 1957 Convention, Art.2.

14 The 1957 convention, Art.3.

15 Griggs, Patrick/Williams, Richard, *Limitation of Liability for Maritime Claims* (1998), p. 3.

The 1976 LLMC was the third international limitation convention for maritime claims.[16] It entered into force in 1986. The claims were also divided into two main categories, i.e. loss of life or personal injury and property claims.[17] The limit of liability for claims covered was raised considerably as well;[18] it was, nevertheless, possible to "strike a balance between successful claimants and shipowners".[19] The 1976 LLMC achieved an increase in the limitation fund to such a sufficiently high level that claimants could reasonably be compensated, but not so high as to make the shipowners' liability uninsurable.[20] In addition, it follows from Article 4 that a person will not be able to limit liability only if "it is proved that the loss resulted from his personal act or omission, committed with the intent to cause such a loss, or recklessly and with knowledge that such loss would probably result".[21] The 1976 LLMC, in effect, provides for a virtually unbreakable system of limiting liability.

In order to enhance compensation and establish a simplified procedure for updating the limitation amounts,[22] the 1996 Protocol to the 1976 LLMC was adopted in London in 1996.[23] Under the said Protocol, the amount of compensation payable in the event of an incident was substantially increased.[24] Moreover, Article 9 of the 1996 Protocol requires, *inter alia*, that, as between the Parties to this Protocol, the 1976 LLMC and the 1996 Protocol shall be read and interpreted together as one single instrument.

Aside from the limitation conventions for maritime claims as discussed above, the limitation system also exists in relation to carriage of goods by sea. It currently comprises the Hague Rules[25], the Hague-Visby Rules[26] and the Hamburg Rules.[27]

---

16 The 1976 LLMC, reprinted in: *Lloyd's Shipping Law Library: The Ratification of Maritime Conventions* (2004), Vol. 2, II.2.330.

17 The 1976 LLMC, Art. 2.

18 The discussion about the amount of the limitation fund will be explained in Section C.II in this chapter.

19 Mandaraka-Sheppard, Aleka, *Modern Admiralty Law* (2001), p. 880.

20 *Ibid.*

21 The 1976 LLMC, Art.4.

22 See the preamble to the Protocol of 1996 to amend the Convention on Limitation of Liability for Maritime Claims of 19 November 1976.

23 The LLMC 1996 Protocol, reprinted in: *Lloyd's Shipping Law Library: The Ratification of Maritime Conventions* (2004), Vol. 2, II.2.340.

24 The LLMC 1996 Protocol came into force in May, 2004. Accordingly, claimants have a much better chance of receiving full compensation for claims arising from shipping incidents. The limits on compensation will increase six-fold for the smallest ships (300-500 gross tons) and by an average of some 250% for other ships. The detailed information about the amount of the funds in the 1996 Protocol will be displayed in Section C.II in this chapter.

25 The Hague Rules were adopted at a diplomatic conference in Brussels in 1924 and were quickly adopted into the municipal legislation of many countries, reprinted in: *Lloyd's Shipping Law Library: The Ratification of Maritime Conventions* (2004), Vol. 3, I.5.10.

26 The Hague-Visby Rules were produced in 1968. These Rules have been adopted relatively quickly into the municipal legislation of many countries, reprinted in:

These rules define the rights and liabilities of the carrier and the cargo interests in an agreement to carry goods by sea and at the same time provide for limitation of the carrier's liability for loss of or damage to cargo.[28] The right of limitation is restricted to claims for the loss or damage incurred in connection with the goods which are being carried. It should also be pointed out that the right to limit for consequential loss can only exist if there is also a claim for physical loss of or damage to the cargo.[29] Whether claims which qualify for limitation under the above rules also qualify for limitation under the 1976 LLMC depends on whether a particular claim is the claim within the meaning of Articles 2 (1)(a) and 2 (1)(b) of the 1976 LLMC.[30]

There are also conventions in relation to the limitation of liability rule in specific areas. For example: the International Convention on Civil Liability for Oil Pollution Damage in 1969 and its 1992 Protocol[31]; the Convention relating to Civil Liability in the Field of Maritime Carriage of Nuclear Material, 1971;[32] the Athens Convention relating to the Carriage of Passengers and their Luggage by Sea 1974 and its 1976 and 2002 Protocols;[33] the International Convention on Liability and Compensation for Damage in Connection with Carriage of Hazardous and Noxious Substances by Sea, 1996; [34] and the International Convention on Civil Liability for Bunker Oil Pollution Damage 2001. Not all of the above conventions have come into force; the latter two and the 2002 Protocol to the Athens Convention relating to the Carriage of Passengers and their Luggage by Sea 1973 are not yet in force. Moreover, it is important to note that the

---

Lloyd's Shipping Law Library: The Ratification of Maritime Conventions (2004), Vol. 2, II.5.20.

27 The Hamburg Rules as set out in the United Nations Convention on the Carriage of Goods by Sea 1978 came into force internationally on 1 November 1992, reprinted in: *Lloyd's Shipping Law Library: The Ratification of Maritime Conventions* (2004), Vol. 2, II.5.220.

28 The limits of liability are provided in the Hague-Visby Rule, Art. IV.5 and the Hamburg Rules, Art.6.

29 Griggs, Patrick/Williams, Richard, *supra,* note 15, p. 113.

30 Article 2(1)(a) and Article 2 (1)(b) read: "1. Subject to Article 3 and 4 the following claims, whatever the basis of liability may be, shall be subject to limitation of liability: (a) claims in respect of loss of life or personal injury or loss of or damage to property (including damage to harbour work, basins and waterways and aids to navigation), occurring on board or in direct connexion with the operation of the ship or with salvage operations, and consequential loss resulting therefrom; (b) claims in respect of loss resulting from delay in the carriage by sea of cargo, passengers or their luggage;..."

31 See Chapter 1, Section A.II.

32 The convention is reprinted in: *Lloyd's Shipping Law Library: The Ratification of Maritime Conventions* (2004), Vol. 3, II.5.170.

33 These conventions are reprinted in: *Lloyd's Shipping Law Library: The Ratification of Maritime Conventions* (2004), Vol. 3, II.5.190, II.5.200 and II.202.

34 The convention is reprinted in: *Lloyd's Shipping Law Library: The Ratification of Maritime Conventions* (2004), Vol. 3, II.7.125.

conventions containing different liability limitation regimes have been adopted for tanker-oil pollution and bunker-oil pollution.

Similar to other issues, any issue regarding the limitation of liability has to take account of the relevant national legislation, which gives domestic effect to these international conventions. International conventions are not independent and they need to be adopted and become a part of the national law of participating countries before they become effective.[35]

II. The 1957 Convention, 1976 LLMC and its 1996 Protocol relating to tanker-oil pollution liability

Under the 1957 Convention, pollution damage could in general come under the category of damage to property, but the said convention did not specify claims for oil-pollution damage. As a matter of fact, since the 1957 Convention did not come into force until 1968, it could not apply to relevant liability issues in the *Torrey Canyon* incident in 1967. This was why that the international community did not wait for the ratification of the 1957 Convention. The discussion for a separate limitation regime regarding oil-pollution liability was immediately launched. Consequently, the 1969 CLC, which contains a set of specific rules for liability and its limitation for tanker-oil pollution damage, was established. The establishment of the 1969 CLC was also due to the development of international shipping, especially the carriage of oil by sea.

Additionally, even if the 1957 Convention had been applicable to the *Torrey Canyon* incident, the compensation amount available under it would have been inadequate for claims arising from the said incident. On the basis of 1,000 gold francs per gross ton as established under the 1957 Convention,[36] the total amount available for compensation would have been FFr.18 million. However, the cost incurred by the British and French governments for the clean-up in the *Torrey Canyon* case was said to be US$16 million, which was around 80 million French

---

[35] Griggs, Patrick/Williams, Richard, *supra*, note 15, p. 113.

[36] The 1957 Convention, Art.3 (1): "1. The amount to which the owner of a ship may limit his liability under Article 1 shall be: (a) where the occurrence has only given rise to property claims, an aggregate amount of 1,000 francs for each ton of the ship's tonnage; ... (c) where the occurrence has given rise both to personal claims and property claims, an aggregate amount of 3,100 francs for each ton of the ship's tonnage, of which a first portion amounting to 2,100 francs for each ton of the ship's tonnage shall be exclusively appropriated to the payment of personal claims and of which a second portion amount to 1,000 francs for each ton of the ship's tonnage shall be appropriated to the payment of property claims; provided however that in cases where the first portion is insufficient to pay the personal claims in full, the unpaid balance of such claims shall rank rateably with the property claims for payment against the second portion of the fund."

francs at the time.[37] It turned out that the compensation amount contained in the 1957 Convention was not sufficient for a major oil-pollution incident.[38]

After the ratification of the 1957 Limitation Convention, some claims against oil-pollution damage could be subject to limitation under the 1957 Limitation Convention. Therefore, the adoption of the 1969 CLC created inconsistencies, since it provided for a set of separate liability and limitation rules for oil pollution damage. To solve this issue, Article 3(b) of the 1976 LLMC thus expressly excludes "claims for oil pollution damage within the meaning of the International Convention on Civil Liability for Oil Pollution Damage, dated 29 November 1969 or of any amendment or Protocol thereto which is in force."[39] The LLMC Protocol 1996 retains the principle of excluding oil pollution from its scope of application. Therefore, if a claim is by nature an oil-pollution claim as defined in the Civil Liability Convention, which in effect refers to the 1969 CLC and its 1992 Protocol, it will be excluded under the 1976 LLMC and its Protocol. Moreover, if the damage is within the meaning of the civil liability convention, even if the State is not a party to the civil liability convention, the 1976 LLMC is not applicable.[40] Meanwhile, the State Parties to the 1976 LLMC are not restricted in their ability to introduce their own limitation legislation in respect of claims for oil-pollution damage.[41]

In the case of an incident involving tanker-oil pollution, two funds subject to different limitation regimes may need to be established. One is for oil-pollution liability and another answers for all other maritime claims, which include property claims and personal injury claims in this incident. These two funds are constituted in accordance with two limitation regimes: the 1976 LLMC or its 1996 Protocol and the CLCs.

---

37 Wu, Chao, *supra*, note 10, p. 34, footnotes omitted.

38 Dykes, Andrew, 'Limitation and Oil Pollution', in: *supra,* note 5, pp. 144-151, at 144: "the emergency of supertankers in the late 1960's, coupled with concern over the potential environmental consequences of casualties involving such ship, led to a widespread desire for an agreed international standard for liability together with ample compensation to be provided for the victims. It was felt that the existing provisions of the 1957 Limitation Convention did not provide an adequate level of compensation; if claims for oil pollution damage were pooled within that Convention's limitation fund, the compensation available for pollution victims might be arbitrarily reduced, depending on the circumstances of the casualty and the size of other claims against the shipowner which would rank ratably against the limitation fund."

39 The 1976 LLMC, Art.3(b): "The rule of this Convention shall not apply to: ... (b) claims for oil pollution damage within the meaning of the International Convention on Civil Liability for Oil Pollution Damage, dated Nov. 29, 1969 or of any amendment or Protocol thereto which is in force;..."

40 Özçayir, Z.Oya: 'Limitation of liability problems in cases of oil pollution', available at: <http://www.turkishpilots.org/DOCUMENTS/Oya_Ozcayir_Liability.htm> (visited 8 June 2005).

41 Dykes, Andrew, *supra,* note 38, pp. 144-151, at 149. That even means that they may ratify the CLC and even extend relevant CLC provisions to bunker-oil spill, or it can have no limitation regime or its separate limitation regime.

The above examination shows that, despite the existence of general limitation conventions for maritime claims, there is always a need for a separate limitation regime for specific shipping matters. The liability for tanker-oil pollution damage is limited under the CLCs instead of under general limitation conventions. This splitting of risks and the creation of separate funds for different types of risks is beneficial to victims, but it may be highly undesirable from the industry's point of view, since it may lead to a number of separate exposures arising out of the same incident.

C. Limitation rules in the Bunkers Convention

The shipowner and the person or persons providing insurance or other financial security have the right to limit liability under the Bunkers Convention.[42] Accordingly, all persons falling within the definition of the shipowner are entitled to limit their liabilities.[43] Moreover, if the persons, except the registered owner, take out insurance voluntarily, their liability insurers can also enjoy the right to limit liability.

Other aspects in respect of the limitation of liability, such as the limitation amount, conditions of limitation and constitution and distribution of the limitation fund are to be regulated "under any applicable national or international regime."[44] The phrase of "any applicable national or international regime" may include: (1) the 1957 Convention; (2) the 1976 LLMC; (3) the 1996 Protocol to 1976 LLMC; and (4) other applicable limitation regimes.[45] The Member States to the Bunkers Convention will have varied limitation regimes and some States may even have no rules of limitation of liability in this respect.[46] This can create confusion in a particular case.

It is widely assumed that the liability for pollution damage caused by bunker spills may be limited under the 1976 LLMC in most jurisdictions where the Bunkers Convention is also in force. This is first due to the fact that the 1976 LLMC entered into force on December 1, 1986 and holds 49.65% of world tonnage with currently 50 Contracting States.[47] Secondly, as mentioned, the resolution adopted following the adoption of the Bunkers Convention will also

42 The Bunkers Convention, Art.6.

43 The Bunkers Convention, Art.1(3).

44 The Bunkers Convention, Art.6.

45 Neither 1976 LLMC nor its Protocol excludes claims for bunker-oil pollution damage not related to tankers. Therefore, if a specific limitation regime had been written into the Bunkers Convention, it would be problematic, since there would be an overlap as regards limitation rules between the 1976 LLMC or its Protocol and the Bunkers Convention.

46 Wu, Chao, 'Liability and Compensation for Bunker Pollution', 33 *J. Mar. L. & Com.* 553 (2002), pp. 553-567, at 562.

47 The information is from: <http://www.imo.org/Conventions/mainframe.asp?topic_id=247> (visited 3 January 2006).

prompt the States to ratify the 1996 Protocol to 1976 LLMC.[48] If the liability for bunker oil pollution is limited under the 1976 LLMC or its 1996 Protocol, the shipowner will not need to constitute a free-standing limitation fund exclusively available for satisfying claims arising from bunker-oil spill incidents.

However, even if all the States involved in a bunker-oil spill incident were Contracting States to the 1976 LLMC or its 1996 Protocol (LLMCs), other problems would arise due to the absence of a specific limitation regime. Two most important issues are: first, would all types of pollution damage resulting from a bunker-oil spill be covered by the LLMCs? Secondly, would the LLMCs provide for adequate funds for the compensation of pollution damage? The following sections will be devoted to examining these two issues.

I. Pollution damage eligible for limitation

1. Pollution damage arising from a bunker-oil spill

The definition of "pollution damage" in the Bunkers Convention is similar to the one in the CLCs.[49] Four types of "pollution damage" are admitted under the CLCs.[50] The first type covers the expenses incurred for preventive measures, including clean-up measures. The liability is for any reasonable measures to prevent or minimise pollution damage: for example, the costs of removing oil (cargo or fuel) from a damaged tanker, as well as the costs of clean-up operations at sea, in coastal waters, on the beach and of disposing of oily wrecks. The claims resulting from the impairment of property, as a second type of pollution damage, include the costs of cleaning contaminated fishing gear, marine-culture facilities, yachts and industrial water intakes. In cases of severe contamination of fishing gear where effective cleaning is impossible, the replacement of the damaged property may be justified. Pure economic loss is the third type of pollution damage. Oil spills can result in economic losses to certain groups such as fishermen and hotel owners. As long as such losses result from oil contamination, the compensation is paid, whether or not the claimant has suffered any damage to his own property. The last category of pollution damage is environmental damage. The compensation is available for the costs of reasonable measures to help restore or reinstate the marine eco-system damaged as a result of an oil spill. The cost of studies related to such measures may also be compensated. Due to the similarity in the definition, a bunker-oil spill incident may also have any type of the above-mentioned pollution damage.

Claims for pollution damage arising out of a bunker-oil spill incident shall be subject to limitation of liability. If the liability is said to be limited under the 1976

---

48 See Chapter 1, Section C.VIII.

49 The Bunkers Convention, Art.1(9), the CLC Protocol 1992, Art. 2(3). However, the types of oil which cause pollution damage are different.

50 The list of the four types of pollution damage is from ITOPF: Oil Spill Compensation: a brief guide to the civil liability and fund conventions, pp. 2-3, available at: <http://www.itopf.com/compensation02.PDF> (visited 8 June 2005).

LLMC or its 1996 Protocol, it is important to examine whether or not any of those claims is subject to limitation.

2. The claims subject to limitation under the 1976 LLMC and its Protocol

The provisions on "Claims subject to limitation" and "Claims excepted from limitation" are nearly the same in the 1976 LLMC and its 1996 Protocol.[51] There are six categories of claims listed in Article 2 of the 1976 LLMC as being subject to limitation of liability. The full text of Article 2 on "Claims subject to limitation" is as follows:

> "1. Subject to Arts. 3 and 4 the following claims, whatever the basis of liability may be, shall be subject to limitation of liability:
> (a) claims in respect of loss of life or personal injury or loss of or damage to property (including damage to harbour works, basins and waterways and aids to navigation), occurring on board or in direct connexion with the operation of the ship or with salvage operations, and consequential loss resulting therefrom;
> (b) claims in respect of loss resulting from delay in the carriage by sea of cargo, passengers or their luggage;
> (c) claims in respect of other loss resulting from infringement of rights other than contractual rights, occurring in direct connexion with the operation of the ship or salvage operations;
> (d) claims in respect of the raising, removal, destruction or the rendering harmless of a ship which is sunk, wrecked, stranded or abandoned, including anything that is or has been on board such ship;
> (e) claims in respect of the removal, destruction or the rendering harmless of the cargo of the ship;
> (f) claims of a person other than the person liable in respect of measures taken in order to avert or minimize loss for which the person liable may limit his liability in accordance with this Convention, and further loss caused by such measures.
> 2. Claims set out in paragraph 1 shall be subject to limitation of liability even if brought by way of recourse or for indemnity under a contract or otherwise. However, claims set out under paragraph 1(d), (e) and (f) shall not be subject to limitation of liability to the extent tat they relate to remuneration under a contract with the person liable."[52]

The 1976 LLMC establishes a regime of limitation for maritime claims; it does not contain any rule in relation to the basis of liability. Therefore, it is irrelevant whether the liability for a bunker-oil spill is strict or not. The above-mentioned Article 2 does not specifically confer the right to limit for claims for bunker-oil pollution damage, and so it is necessary to examine whether any pollution damage

---

51 There is a small difference. For instance, Art. 3 (1) of the 1976 LLMC provides that: "The rules of this Convention shall not apply to: (a) claims for salvage or contribution in general average;..." In the 1996 LLMC, it is more specifically provided, in Art. 3(a), that "The rules of this Convention shall not apply to: (a) claims for salvage, including, if applicable, any claim for special compensation under Article 14 of the International Convention on Salvage 1989, as amended, or contribution in general average;..."

52 The 1976 LLMC, Art. 2.

caused by a bunker-oil spill incident falls within one of the above six categories of claims in order to be subject to limitation under the LLMCs.

Article 2(1)(a), as the paragraph prescribing the first category of claims, includes the loss of life, personal injury claims and also property claims including consequential loss. As far as pollution damage is concerned, the loss of life and personal injury claims are irrelevant. The claims for "loss of or damage to property", however, must be for one that occurred "on board or in direct connexion with the operation of the ship, or with salvage operations, and consequential loss resulting therefrom." The 1976 LLMC does not give further definition of the relevant wordings such as "on board" or "the operation of the ship". The leading case on this subject that has often been cited is the *Tojo Maru*[53] case, which was decided in the UK House of Lords. In that case, a diver, operating from a salvage tug, caused serious damage to the vessel on which the salvage work was being carried out. The tug owners could not limit their liability, since it could not be said either that the diver was on board the tug or the vessel to which salvage services were being rendered or that he was acting in a way which was part of "management or navigation" of the salvage tug.[54] In addition, the term "operation" is broad, as it may be held to relate to its physical use or its purely commercial use or operation.[55] Therefore, it is true that "...limitation was restricted to acts or omissions done by a person on board or in the navigation or management of the ship, or in the loading, carriage or discharge of its cargo, or in the embarkation, carriage or disembarkation of its passengers."[56] This category furthermore extends to "consequential loss resulting therefrom". Accordingly, if "economic loss" is consequential upon the damaged property, the claim in this respect falls within this category. However, the "pure economic loss" such as the loss of income sustained by fishermen and hotel owners after a bunker-oil spill may not be covered under this paragraph, since there is no damage to property.[57]

Article 2(1)(b) provides the right to limit for "claims in respect of loss resulting from delay in the carriage by sea of cargo, passengers or their luggage." The limitation rule in the 1976 LLMC thus applies to the loss resulting from the delay of the carriage. There is no further definition of "loss". Suppose, for instance, that a cargo ship was delayed due to the fact that bunker oil had been spilt, would all claims in respect of the losses fall within this article? If the cargo ship had cargo on board, three different situations could be taken as examples: first, the cargo was contaminated by the spilt oil and the loss was caused during the undertaking of the clean-up measures; secondly, the cargo was a type of normal cargo, but after waiting for the cleaning-up work, the cargo-owner suffered consequential financial losses due to market fluctuations; thirdly, the cargo was perishable in nature and after the clean-up effort, the cargo lost its value or the price had to be lowered

[53] [1971] 1 Lloyd's Rep. 341.
[54] Grime, R.P., *supra*, note 10, at 309.
[55] Geoffrey Brice, Q.C., 'The Scope of the Limitation Action', in: *supra*, note 5, pp. 18-32, at 23.
[56] Griggs, Patrick/Williams, Richard, *supra*, note 15, p. 16.
[57] See IMO LEG 74/4/2.

because of the devaluation. These three categories of loss might fall within the "loss resulting from delay". Meanwhile, it is also interesting to examine whether the "loss" are "pollution damage" covered by the Bunkers Convention. As seen in the first case, pollution damage caused to the cargo was inside the ship. It was not the "loss or damage caused outside the ship by contamination" as required by the Bunkers Convention.[58] Furthermore, in the second and third case, the "pure economic loss" in the sense of "pollution damage" under the Bunkers Convention means the loss of earnings as a result of the contaminated environment. The loss in the said examples, however, resulted from the market fluctuation or the devaluation of the cargo itself. Apparently, there is a need to differentiate between pollution damage claims and general maritime claims in some circumstances. However, if the loss resulted from the delay, it does not affect the right of the shipowner to limit his liability under the 1976 LLMC in any way.

Article 2(1)(c) is a provision which was intended to be a residual category designed to sweep up liabilities that might be imposed on shipowners.[59] It allows the benefit of limitation where rights have been infringed. The wording of "other loss" in this paragraph indicates losses other than those referred to in the two preceding sub-paragraphs (a) and (b).[60] The true meaning of the terminology of infringement of rights is not really clear.[61] The right infringed is, however, not intended to cover contractual rights, but there is no further description of the scope of non-contractual rights falling within the said paragraph. In effect, this paragraph is difficult to apply in practice. For instance, when a bunker-oil incident occurs, the reduction in tourism following an oil spill that pollutes the beaches may lead to a loss of earnings for hotel owners, restaurant owners, etc. Would these losses fall within Article 2(1)(c)? The answer depends on the interpretation of "the infringement of rights other than contractual rights". Since there is no further explanation in the 1976 LLMC, it depends on national legislation.

Article 2(1)(d) grants a right to limit liability for "claims in respect of the raising, removal, destruction or the rendering harmless of a ship which is sunk, wrecked, stranded or abandoned, including anything that is or has been on board such a ship." Meanwhile, Article 2(2) is to the effect that claims falling with this paragraph shall not be subject to limitation if they relate to any remuneration under a contract with the person liable.[62] Accordingly, in a bunker-oil spill, the clean-up costs and expenses for removal measures will be covered by this category if the ship was sunk, wrecked, stranded or abandoned. However, the limitation rule would not be applicable if the condition was not met, as, for instance, in a collision that caused a spill, but did not result in the ship being wrecked.[63] In addition, in accordance with Article 18(1) of the 1976 LLMC, which provides that "Any State may, at the time of signature, ratification,

[58] The Bunkers Convention, Art. 9(a).
[59] Grime, R.P., *supra*, note 10, at 310.
[60] Geoffrey Brice, Q.C., *supra*, note 55, at 24.
[61] Chen, Xia, *Limitation of Liability for Maritime Claims* (2001), p. 38.
[62] The 1976 LLMC, Art.2(2).
[63] See IMO LEG 74/4/2.

acceptance, approval or accession, reserve the right to exclude the application of Article 2 paragraph 1(d) and (e)..."[64], States accordingly have the right of reservation and avoiding the application of limitation of liability in wreck-removal claims. It may occur that States subject claims for wreck-removal costs to unlimited liability.

The paragraph (e) confers the benefit of limitation where claims relate to the removal, destruction or the rendering harmless of the cargo of the ship. It has been pointed out that first, there is a conflict between Article 2(1)(d) and Article 2(1)(e)[65], since "the cargo of the ship" in Article 2(1)(e) may fall within the scope of "anything that is or has been on board such ship" in Article 2(1)(d).[66] Secondly, in the same vein under Article 18(1), contracting States can also exercise the right of reservation and avoid paragraph (e). It may happen that contracting States only make a reservation regarding one of these two paragraphs. For instance, the United Kingdom has made a reservation to Article 2(1)(d) but not to Article 2(1)(e); consequently, claims concerning cargo removal qualify for limitation before the ship is sunk, wrecked, stranded or abandoned but not after that event has occurred.[67]

Paragraph (f) stipulates that claims in respect of the costs of the action taken to minimise loss are subject to limitation. In an oil-spill incident, it is rather frequent that preventive measures are taken to prevent or minimise pollution damage. Claims have to be made by a person other than the person liable. In addition, the "loss" must be the loss for which the liable person is liable. However, in an incident such as an oil spill, the shipowner, as the liable person, is also most likely involved in averting or minimising pollution damage, and in fact he is always in the forefront position to eliminate or prevent pollution damage. The loss he may suffer in the process of preventing or minimising pollution damage cannot, however, fall within this paragraph.

To sum up, the Bunkers Convention's attempt to put limitation of liability issues under the LLMCs will not be entirely successful. Some claims for pollution damage caused by a bunker-oil spill cannot be eligible for limitation under the 1976 LLMC: for instance, Article 2 of the 1976 LLMC does not allocate any sub-paragraph for environmental damage.

II. The amount of the funds available under the LLMCs

In a bunker-oil spill incident, claims for pollution damage and other maritime claims may be filed at the same time. If the 1976 LLMC or its Protocol is applicable, both types of claims share one limitation fund. As such, one concern is that claims for pollution damage may receive less payment than if a separate limitation fund existed for it.

64 The Bunkers Convention, Art. 18(1).
65 Griggs, Patrick/Williams, Richard, *supra*, note 15, p. 18.
66 Chen, Xia, *supra*, note 61, p. 47-48.
67 Griggs, Patrick/Williams, Richard, *supra*, note 15, p. 18.

The applicable limitation regime is varied by virtue of Article 6 of the Bunkers Convention.[68] However, the guaranteed amount of compensation from the liability insurer is in all cases "not exceeding an amount calculated in accordance with the Convention on Limitation of Liability for Maritime Claims, 1976, as amended."[69] The limitation amounts in the 1976 LLMC and its 1996 Protocol are thus chosen to display the possible amount of the fund available for pollution victims. Loss of life or personal injury claims will not be considered in this context.

In accordance with Article 6(1) of the 1976 LLMC, the limit of liability for claims other than loss of life or personal injury is fixed at 167,000 SDR for ships not exceeding 500 tons. For larger ships, the additional amounts are: (1) for each ton from 501 to 30,000 tons, 167 SDR; (2) for each ton from 30,001 to 70,000 tons, 125 SDR; (3) for each ton in excess of 70,000 tons, 83 SDR.[70]

It was the purpose of the LLMC Protocol 1996 to increase the limit. The limit of liability for property claims for ships not exceeding 2,000 gross tons is pegged at 1 million SDR instead of 417,500 SDR under the 1976 LLMC. For larger ships, the following additional amounts are used in calculating the limitation amount: (1) for each ton from 2,001 to 30,000 tons, 400 SDR; (2) for each ton from 30,001 to 70,000 tons, 300 SDR; (3) for each ton in excess of 70,000, 200 SDR.[71]

The amounts mentioned above shall be converted into the national currency of the State in which limitation is sought, according to the value of that currency at the date the limitation fund shall have been constituted, payment is made, or security is given which under the law of that State is equivalent to such payment.[72] The value of the national currency in terms of the SDR is specially classified in two groups: if a State Party is a member of the International Monetary Fund, the value of a national currency in terms of the SDR shall be calculated in accordance with the method of valuation applied by the International Monetary Fund in effect at the date in question for its operation or transaction. However, if the State is not a member of the International Monetary Fund, the value of a national currency in terms of the SDR shall be calculated in a manner determined by that State Party.[73]

An "ideal" situation in the case of a bunker-oil spill is when it results exclusively in bunker-oil pollution damage and all the States involved are Member States of the LLMC Protocol 1996. The victims would thus benefit from a specific and uniform limitation regime. However, the adequacy of the fund to compensate all pollution victims in this case is still questionable. For instance, the compensation amount available is SDR 3 million under the CLC Protocol 1992 if a tanker of 5,000 gross tons is involved. By contrast, a claim against a non-tanker ship of 5,000 gross tons can only amount to 2.2 million SDR under the LLMC

68 The Bunkers Convention, Art. 6.

69 The Bunkers Convention, Art. 7(1).

70 See 1976 LLMC, Art. 6(1)(b). The value of SDR is determined daily by the International Monetary Fund on the basis of a basket of currencies; the limitation amounts can be assessed and calculated.

71 The LLMC Protocol 1996, Art.3.

72 The 1976 LLMC, Art.8 (1).

73 The 1976 LLMC, Art.8(1).

Protocol 1996.[74] Moreover, the factual situation is far more complicated once States involved in an incident are contracting States to different limitation regimes.

Shipowners as well as claimants are always keen to know the scope and extent of the liability. Under the CLCs, claimants are informed of the maximum recovery which they can expect from the shipowner due to the fact that the shipowner's liability is limited to a known figure. It is a success of the CLCs to have a separate limitation regime instead of being put alongside with other claims that may arise from a marine incident. Comparatively, the applicable limitation of liability rule for a bunker-oil spill will depend on the facts of the case, as does the amount of the fund.

III. Other aspects relevant to claims for bunker-oil spill liability under the 1976 LLMC and its Protocol

1. *Conduct barring the right to limit*

Article 4 of the 1976 LLMC describes conduct barring the shipowner's right to limitation as follows:

> "A person liable shall not be entitled to limit his liability if it is proved that the loss resulted from his personal act or omission, committed with the intent to cause such loss, or recklessly and with knowledge that such loss would probably result."[75]

The LLMC Protocol 1996 follows the same approach.[76]

It is fairly reasonable that the shipowner shall be liable for loss of or damage to others caused by his fault. By virtue of this Article, it is not easy to overturn the shipowner's right to limit his liability. The provision first requires that the loss results from a "personal" act or omission. The "personal" act or omission includes an act or omission by the following persons: the shipowner, the charterer, manager, operator, salvor, liability insurer of the vessel or "any person for whose act, neglect or default the shipowner or salvor is responsible"[77] Due to the effect of the insertion of "personal", it has been analysed that, presumably, the personal act of any one identified in Article 1 of the LLMC will prevent him from limiting his own liability in the event of a claim against him but will not necessarily defeat the right to limit of any other persons in the same group in the event of a claim against them.[78] The approach appears to be similar in the Bunkers Convention. In Article 7(10), it provides that "...even if the shipowner is not entitled to limitation of liability according to Article 6, the defendant may limit liability to an amount equal to the amount of the insurance or other financial security required to be

[74] The amount is calculated according to Art.3 of the LLMC 1996 Protocol.
[75] The 1976 LLMC, Art. 4.
[76] It is also the same in the CLC Protocol 1992.
[77] The 1976 LLMC, Art. 1.
[78] Griggs, Patrick/Williams, Richard, *supra*, note 15, p. 28.

maintained in accordance with the paragraph 1…"[79] The defendant in this article refers to the liability insurer. It means that the liability insurer has the right to limit his liability even if the shipowner is not aimed at protect the interests of the liability insurer. Article 7(10) of the Bunkers Convention protects the interests of the liability insurer.

At the same time, the requirement of "with the intent…or recklessly…" is difficult to prove since: (1) it specifies in precise terms the mental element which must be proved of the person entitled to limit; (2) the mental element of "intention or recklessness" relates to the consequence and not to the action itself. It is possible that a person may intend to be reckless about the act but is legitimately neutral about the consequence of that act.[80] Therefore, it may be regarded as narrowing the circumstances in which the right to limit may be lost.[81]

2. Constitution and distribution of the limitation fund

Under the CLCs, the owner, for the purpose of availing himself of the benefit of limitation, shall constitute a fund for the total sum representing the limit of his liability with the court or other competent authority.[82] The fund can be constituted either by depositing the sum or by producing a bank guarantee or other guarantee, acceptable under the legislation of the Contracting State where the fund is constituted.[83] In practice, it is often desirable for a liable person seeking to limit his liability to set up a limitation fund at an early stage, since upon the establishment of a limitation fund, claimants can be prevented from claiming against other assets belonging to the liable person.

Article 10 of the 1976 LLMC allows a liable person to invoke the right to limit his liability without having constituted a limitation fund, although a State Party may opt to provide in its national law that a person liable may only be entitled to limit his liability if a limitation fund has been constituted.[84] Some State Parties have chosen not to place restrictions on a person's right to invoke limitation of liability without having constituted a fund: for instance, the United Kingdom.[85]

From Article 11 to Article 14, the 1976 LLMC contains detailed provisions regarding the constitution and distribution of a limitation fund. Briefly, according to Article 11, the shipowner may constitute a fund, either by depositing the sum or by producing a guarantee acceptable under the legislation of the State Party where the fund is constituted and considered to be adequate by the court or other competent authority.[86] Meanwhile, any fund thus constituted shall be available only for

---

[79] The Bunkers Convention, Art.7(10).

[80] Grime, Robert, 'The Loss of the Right to Limit', in: *supra*, note 5, pp. 102-112, at 100.

[81] *Ibid.*

[82] The CLCs, Art. V(3).

[83] *Ibid.*

[84] The 1976 LLMC, Art.10.

[85] Patrick Griggs, *supra*, note 15, p. 46.

[86] The 1976 LLMC, Art.11(2).

the payment of claims in respect of which limitation of liability can be invoked.[87] Furthermore, by virtue of Article 13, if the shipowner is entitled to limitation, the constitution of a limitation fund will reassure the shipowner that his other assets will be protected from the claimants.[88]

The established fund should be distributed proportionately among the claimants according to their established claims.[89] The rules for distribution of the fund are provided in Article 12 of the 1976 LLMC. The same article grants the shipowner or his liability insurer the right of subrogation against the fund if he settles claims before the fund is distributed.[90]

Liability claims against bunker-oil pollution damage are not an independent class of claims under the 1976 LLMC, and so pollution victims who suffered in a bunker-oil spill do not have a privilege over other claims.

D. The right to limit liability

I. The reasons for maintaining the right to limit

The limitation of liability is a long-standing principle in maritime law. Although legal provisions regarding limitation of specific liability vary with time and place, it enables the shipowner – where applicable – to limit his liability for the loss or damage arising from the operation of the ship.

The traditional reason for having a limitation system for the shipping industry was mainly related to the nature and extent of the risks involved in maritime transportation, since it was a very risky business.[91] The right to limit liability was intended to attract people to invest in the shipping industry and so build up a competitive mercantile marine.[92] With the development of the shipping industry

---

[87] The 1976 LLMC, Art.11(1).

[88] The 1976 LLMC, Art.13(1).

[89] The 1976 LLMC, Art.12(1).

[90] The 1976 LLMC, Art.12(2), see also Art.12(3): "The right of subrogation provided for in paragraph 2 may also be exercised by persons other than those therein mentioned in respect of any amount for compensation which they may have paid, but only to the extent that such subrogation is permitted under the applicable national law."

[91] Beddard, Ralph, *supra*, note 6, at 152. Chen, Xia, *supra*, note 61, p. xiii: "A shipowner is not only exposed to the perils of the sea, but also vulnerable to the negligence of the master and crew members under the doctrine of *respondeat superior*, which holds the shipowner vicariously liable for the negligence of his employees."

[92] Gauci, Gotthard, 'Limitation of Liability in Maritime Law: an Anachronism?' 19 *Marine Policy* 65 (1995), pp. 65-74, at 66. See also Geoffrey Brice, Q.C., *supra*, note 55, at 18: "its origins are to be found in the desire of governments and the legislature to protect the financial interests of shipowners so as not to discourage the ownership and operation of ships."

and relevant technology, this reason for the limitation of liability is, however, no longer convincing today.

Some other arguments remain or appear to support the right of the shipowner to limit his liability. First, it is preferred by the shipping and cargo industry that the maximum amount of liability under a limitation regime be known in advance, which can facilitate the arrangement of their business. Secondly, the limitation of liability provides a mechanism through which all the costs consequent on a maritime catastrophe can be shared between all those involved in and benefiting from the maritime adventure.[93] In particular, a certain number of incidents are unavoidable and there will seem to be a very strong case for sharing their costs rather than imposing all the losses on the shipowner, particularly in the case where there was no real fault on his part.[94] This idea is in effect applied in the international civil liability system. In that system, the owner of the ship and the cargo share the burden of compensation: this may arise in a tanker-oil spill incident. Thirdly and most distinctively, the shipowner's expanding liability necessitates the insurance arrangement since it is impossible for him to assume all liabilities. At the same time, it is difficult to obtain liability insurance unless a maximum amount of liability is fixed and known in advance.

II. Unsatisfactory outcome of the limitation regime

The applicable limitation rule for bunker-oil pollution liability is not certain under the Bunkers Convention. It may confuse the industries that intend to arrange a carriage, since the question of whether the liability is limited or not will depend on any applicable law. Some delegations to the discussion of a bunkers convention warned that: "…this Article ("Limitation of liability") should contain its own figures and should not refer to other instruments such as LLMC."[95] In their opinion, "Article 6 of the draft convention was vague and confusing."[96] It is even possible that the shipowner's liability will not be limited if no limitation rule is provided for under the applicable national law.[97] Furthermore, since the applicable limitation regime is uncertain, the shipowner and his liability insurer, who have already set up a limitation fund in one country, may later be sued for the payment of further sums in proceedings commenced in some other countries where different limitation rules are applicable.

---

93 Seward, R. C., 'The Insurance Viewpoint', in: *The new law: Limitation of Shipowners' Liability* (1986), *supra*, note 5, pp. 161-186, at 163.

94 *Ibid.*

95 See IMO LEG 76/WP.3.

96 *Ibid.*

97 Tsimplis, Michael N., 'The Bunker Pollution Convention 2001: Completing and Harmonising the Liability Regime for Oil Pollution from Ships?'' 1 *Lloyd's Mar. & Com. L.Q.* (2005), pp. 83-100, at 83.

There certainly appeared to be general agreement that "the limits of liability in the draft bunkers instrument should be tied to those in the LLMC."[98] However, the phrase of "any applicable national or international regime" can be broad enough to contain any applicable limitation regime. The resolution in this regard urges States to ratify the LLMC Protocol 1996. However, the resolution is only a recommendation; States are free not to accept the resolution and maintain their legislation in respect of the limitation of liability even after the ratification of the Bunkers Convention.

One could imagine that it was not easy for the drafters to arrive at the provision as regards limitation of liability in the Bunkers Convention. It was believed that the shipowner and his liability insurer would be better off without another limitation regime. If another limitation fund were required to be established under the Bunkers Convention, it would increase both the shipowner's and the insurer's total exposure.[99] A bunker-spill incident may involve property damage, personal injury and pollution damage, and so the shipowner would need to take out different types of insurance or set up limitation funds in order to cover his potential losses.[100]

However, there are also disadvantages in having such a "one for all" limitation regime. The claims for bunker-oil pollution damage have to be paid for from the same fund as non-pollution claims under any limitation convention. Pollution victims will have to compete with other types of victims in order to get compensation. It is a matter of concern whether the limitation amount will be high enough to satisfy the need for adequate compensation.[101] This will also slow down the compensation procedure for pollution victims. Therefore, this might not satisfy some States that were interested in acceding to a convention intended primarily to protect the interests of pollution victims. It may be important to recall that even a small amount of bunker oil can cause pollution damage which needs a high amount of compensation. Therefore, if the applicable national limitation rule is in

---

98 See IMO LEG 77/11.WPD: "There was general agreement in the Committee that the limits of liability in the draft bunkers instrument should be tied to those in the LLMC, and accordingly no separate limits of liability would be established."

99 Wu, Chao, *supra*, note 46, at 564: "If the Bunkers Convention has instead provided for a dedicated limitation fund, it would have created an additional burden to shipowners/insurers, because, for many cases, at least two limitation funds would have to be established in the event of a bunker spill (76 LLMC fund, BC fund). And if the spill comes from a vessel carrying HNS cargo, then three funds would have to be created (HNS fund, 76 fund, BC fund)."

100 See IMO LEG 74/4/4: "In addition, the Australian government made the observation in its submission to the last session of the Legal Committee (LEG 73/12) that the Convention would need to be closely coordinated with the CLC and HNS Conventions so that double limits would not apply to any one incident, e.g. when bunker oil is spilled from a ship carrying HNS at the same time as some of its cargo is released. A related point is that there may be incidents where the shipowner would be required to constitute three separate funds from the group up from CLC, HNS and bunkers-a potentially expensive exercise and an inefficient use of insurance capacity."

101 Some discussions see Section C.II of this chapter.

favour of the shipowner's interests and has established a low limitation of liability, it will be detrimental to the interests of pollution victims.

It might have been possible to choose to set up a free-standing limitation regime for bunker-oil pollution liability. However, this was not accepted. It is thus necessary to convince State Parties that it is more appropriate that limitation of liability for a bunker-oil spill be fixed in accordance with the general maritime limitation regime, instead of under a separate limitation regime.

E. The relation of limitation of liability and insurance

I. The insurability and limitation of liability

As mentioned above, the most distinctive reason for maintaining the right to limit liability is to make liability insurance available to shipowners. As a matter of fact, the discussions at the 1969 CLC to the Bunkers Convention show that the principle of limitation of liability is considered along with the insurability. For instance, during the preparatory work of the 1969 CLC, the German Maritime Law Association pointed out that: "...the limitation principle must be maintained. This is necessary in view of the enormous amounts that are involved in all maritime adventures and also for the practical purpose of insuring these risks."[102] The British Law Association had almost the same opinion: "...It is for practical purposes essential to the insurance of a liability risk that a maximum liability is fixed and known in advance." [103] During the negotiations of the Bunkers Convention, it was also recognised that a clear right to limit liability was desirable, since once the liability could be limited in accordance with relevant rules, i.e. a regime where the risks can be precisely calculated, the insurance against the risk could be thereafter obtained.

As a general principle, a risk suited for insurance in an ideal situation would meet the following requirements: (i) the potential loss would be significant but the probability would not be high, thus making insurance economically feasible; (ii) the probability of the loss to be insured would be accurately calculated by the

[102] See C.M.I. Documentation III, The German Maritime Law Association, pp. 32-41, at 36.

[103] The British Law Association, see C.M.I. Documentation-III, pp. 50-59, at 56. In C.M.I. Documentation 1968-I &1968-III, Many other countries considered limitation of liability together with insurance. For example, the Belgian Maritime Law Association, see C.M.I. Documentation-III, pp. 98-107, at 104: "what is more, if this limitation did not exist, it would very likely be impossible for the shipowner to find underwriters accepting to cover unlimited liability." The Finnish Maritime Law Association, see C.M.I. Documentation-III, pp. 10-17, at 12: "Insurance is based on the principle that premiums paid or to be paid by the insured in the long run should cover his losses and in addition give the insurer a profit. In the cases now under discussion, the insurance would have to be for an unlimited amount or for an amount whose ceiling is very high. Most individual shipowners in the oil trade would find it difficult to make the necessary insurance arrangement."

prospective insurer; (iii) there would be large numbers of homogeneous exposure units; (iv) the risks assumed by an insurer would involve only the loss which would be accidental and fortuitous; (v) the insurer would be able to learn the time and place of loss in order to determine whether or not the loss were to be covered; (vi) there would be limits beyond which the prospective insurer would be reasonably sure that losses were not to go, i.e. there is no catastrophe exposure.[104] However, it has also been explained by the same author that: "…As a practical matter, many risks which are insured meet these requirements only partially or, with reference to a particular requirement, not at all. Thus, in a sense, the requirements listed described those which would be met by the ideal risk…"[105]

In particular, marine risks are insurable in most cases. It is normal practice for marine insurers to try to ascertain the extent and scope of the risks before they decide to insure. A vessel is insured at a binding valuation in hull insurance since "that valuation provides the most important factor in assessing the cost of the insurance"[106] and "it would seem plausible to argue that the contingent liability of the ship should bear some relation to that value if the hull insurers are to be content to underwrite it."[107] By contrast, the liability cover of the shipowner is predominantly offered by P&I Clubs on mutual insurance terms, "…All that is then left is the general point that a predictable outside limit, by defining the maximum exposure, reduces uncertainty and is cheaper to insure…"[108] The maximum financial exposure of liability insurers might be predicted beforehand[109]: for instance, liability insurers can base their prediction by reference to the age of the vessel, the cargo carried, trading area, the past claims record of the owner and the financial risk by past experience, and so forth. Meanwhile, reinsurance can be arranged above a certain level in order to cover large-scale liability. It is a means for the insurer to relieve himself of the risks that he agreed to undertake.

In practice, P&I Clubs impose limits on shipowners' liabilities only for claims of a catastrophic or "overspill" nature. Oil-pollution damage claims are categorised as a type of "overspill" claim. Although reinsurance is arranged by the International Group of P&I Clubs, due to the disastrous consequences and high costs that will be involved in oil-spill incidents, the limit is imposed on oil-pollution claims. It has been a concern for P&I Clubs that if the liabilities for certain risks such as oil-pollution damage were not limited, the insurance would not be able to afford indemnity in the end. It has been suggested that if the liability is a potentially unlimited liability, the available reinsurance arrangement might even disappear.[110]

---

[104] Athearn, James L., *Risk and Insurance* (1977), pp. 31-36.

[105] See *ibid*., p. 32.

[106] Grime, R.P., *supra*, note 10, p. 308, F.N.10.

[107] *Ibid.*

[108] *Ibid.*

[109] See *ibid*., although "…the concept of 'maximum insurable risk' is very hard to define."

[110] Seward, R.C., 'The Insurance Viewpoint', in: *supra*, note 5, pp. 161-186, at 164: "…the reinsurance available today might not be there in the future in the event of

The P&I Clubs maintain their position of declining to undertake potentially unlimited liability. The required COFR under the OPA 90 is an example in this respect.[111] Under the OPA 90, the exceptions negating the right to limit liability are quite broad.[112] The limitation amounts to be guaranteed are very much in excess of those provided under the CLCs. For example, the COFR guarantee in respect of *Exxon Valdez* under the CLC would be approximately US$8.5 million, but under OPA 90 it would have been US$120 million.[113] It was believed that such a high limit could easily become an unlimited liability.[114] This possibly unlimited liability is one of the significant reasons for insurance companies and P&I Clubs to refuse to be guarantors for the COFR under the OPA 90.[115]

II. The possibility to have a unlimited liability

In the 74th session of the IMO Legal Committee, unlimited liability for bunker-oil pollution damage was proposed by Greenpeace International. They proposed to put unlimited liability on the shipowner and this was intended to cover the full costs of pollution damage resulting from a bunker-fuel oil spill.[116] They believed that the shipowners directly benefit from the risky activity; it is thus appropriate that they assume the full costs of their business activity.[117]

Along with the limitation of the shipowner's liability, the Bunkers Convention does not create a device by which the various parties interested can share liabilities arising from a bunker-oil spill. The owner of the cargo is not required to be liable under the Bunkers Convention.[118] However, the shipowner has the chance to make an arrangement to share potential liabilities with other persons involved in maritime transportation. For instance, the shipowner can raise the price of the charter party in order to share the burden of potentially enormous pollution-damage liability. It might follow that it does not matter whether there is limited or unlimited liability, it is possible for the shipowner to distribute liabilities and minimise his exposure to all liabilities.

In theory, it has been analysed by one author that although it is difficult to insure unlimited liability, it does not mean that the insurance consideration should

---

significantly large claims coupled with the demise of the right to limit..." See also Chapter 4, Section F.I.

111 For more see Chapter 2, Section C.II.2.

112 U.S.C. §§ 2704, see also Chapter 2, Section C.II.2.a).

113 Bryant, Hugh, 'Specialist Insurers Offer Real OPA Solution for Shipowners', *Lloyd's List*, Friday June 21 1996.

114 But see *ibid.*, it says this belief was wrong.

115 For more detailed analysis see Alcantara, Leonard F./Cox, Mary A., 'OPA 90 Certificate of Financial Responsibility', 23 *J. Mar. L. & Com.* 369 (1992), pp 369-386.

116 See IMO LEG 74/4/3.

117 See IMO LEG 74/4/3.

118 See Chapter 4, Section F.II.

determine the whole liability system and its limitation structure. He thinks from the perspective of insurance costs that:

> "The role played by insurance costs in competition seems frequently to have been exaggerated in international discussion, because the introduction of unlimited liability would mean only a marginal (if even that in a long perspective) increase in overall operating costs (of which liability insurance represents only a small portion). Consequently, I cannot accept insurance costs as a key argument for limitation of maritime liability."[119]

The limit of liability, i.e. the capping of catastrophe exposure, can provide a considerable degree of comfort to the insurance industry.[120] In addition, it can speed up the settlement of claims. Accordingly, it ensures both the insurer and the shipowner a greater degree of certainty for their liability. However, this should not be done at the cost of sacrificing the interests of pollution victims. Therefore, it is necessary to strive for a balance "between the desire to ensure on the one hand that a successful claimant should be suitably compensated for any loss or injury which he had suffered and the need on the other hand to allow shipowners, for public policy reasons to limit their liability to an amount which was readily insurable at a reasonable premium".[121]

As far as the liability insurance market is concerned, even if the current insurance market could not cover the whole extent of liability, it would be possible to create new insurance providers. As mentioned in Chapter 2, some specialist insurers have appeared for the purpose of satisfying the COFR requirement under the OPA 90.[122] Two of them are called *Shoreline Mutual* and *the First Line programme of Stockton Re*. Both insurers offer COFR guarantees up to US$395 million for any one ship.[123] The P&I Clubs continue to provide insurance for oil-pollution damage on the US market, which is, however, up to an agreed limit. The emergence of new insurance providers for the COFR requirements shows that there may rarely be a shortage of insurance capacity or insurance providers for high-limit coverage at commercially acceptable levels.

If it is necessary to maintain a right of limitation of liability for liability insurers, this should not mean that the shipowners' liability would be limited. However, neither should it be advisable to establish unlimited liability without insurance arrangements. The most often cited case is where a ship is owned by a single-ship company and its resources consist only of the value of the vessel. In this case, potential victims of a marine pollution incident involving this company would only get that wrecked ship or even nothing. Therefore, unlimited liability without an insurance back-up would not be a good alternative as far as the

---

119 Wetterstein, Peter, 'P&I and Environmental Damage', in the seventh Axel Ax: son Johnson Colloquium on Maritime law Hässelby Colloquium, May, 27-28, 1993: *P&I Insurance*, pp. 115-139, at 134.

120 Steel, David, 'Ships are Different: the Case for Limitation of Liability', *Lloyd's Mar. & Com. L.Q.* (1995), pp. 77-87, at 87.

121 Griggs, Patrick/Williams, Richard, *supra*, note 15, p. 3.

122 More see Chapter 2, section C.II.2.

123 Lloyd's List: *supra*, note 113.

interests of victims were concerned. The best solution might be to impose unlimited liability on the shipowner with compulsory insurance up to a specific amount.[124]

A possible solution would be to limit insurance coverage while leaving the whole liability unlimited. It is said that limited liability insurance coupled with unlimited liability is no longer a matter reserved for academic research. It has been applied in the field of liability for nuclear damage in Swiss law.[125] In addition, in practice, claims against the P&I Club are subject to deductibles, and the Clubs require the members to bear part of the loss[126]; this is one aspect showing the divisible interests between the Club and its members. Furthermore, since the "pay to be paid" rule is somehow abandoned in oil-pollution cases, as can be seen below in Chapter 6,[127] it would be possible to leave the liability unlimited and let the shipowner bear the portion of the liability above the insurance limit alone. Put in another way, we could impose unlimited liability on the shipowner; the liability insurer could contract to cover a limited amount of liability. The right of victims to claim against the shipowner directly above the insurance limit would be retained independently of the Bunkers Convention.[128] Therefore, if the amount which the liability insurer were to pay pollution victims were less than the amount of the shipowner's liability for his victims' compensation, the shipowner would remain liable for the balance. Accordingly, it could be an option for the drafters of the Bunkers Convention to stipulate an unlimited liability for the shipowner with his compulsory insurance up to "an amount calculated in accordance with the Convention on Limitation for Maritime Claims, 1976, as amended". However, unlimited liability was not really taken into account during the preparatory work of a bunkers convention.

---

[124] Gauci, Gotthard, *supra*, note 92, at 67. See also Wetterstein, Peter, *supra*, note 119, p. 135, "It is probable that were unlimited liability introduced, insurers would fix a ceiling on their liability (cf. the current discussion); this would be dependent on the insurance capacity-which would soon adapt to the new situation (there might of course be a temporary decrease in market capacity following a big disaster). The excess, more or less theoretical (cf. the experience so far), liability would then fall upon the shipping industry."

[125] Gauci, Gotthard, *ibid.*, Footnote 30: it has been stated that since 1/1/1984, a nuclear operator is liable in terms of Swiss law without any benefit of limitation of nuclear damages which are occasioned by nuclear materials in his installation; obligatory insurance is limited to a specific amount. See also Faure Michael (ed.), *Deterrence, Insurability, and Compensation in Environmental Liability: Future Developments in the European Union* (2003), p. 201: "...Recent examples have also shown that with respect to the nuclear liability conventions some countries have introduced a duty to insure up to a limited amount, but have left the liability of the licensee of the nuclear power plant itself unlimited."

[126] Brown, Robert H., *Introduction to Marine Insurance: Training Notes for Brokers* (1995), in section 14, p. 5.

[127] See Chapter 6, "pay to be paid" rule is described in Section C.III of Chapter 6.

[128] Gauci, Gotthard, *supra*, note 92, p. 66.

F. Concluding remarks

As a general principle, the shipowner enjoys the right to limit his liability for maritime claims. The system of limitation of liability shall contain a level of limitation high enough to satisfy all genuine claimants. Meanwhile, the prescribed limitation has to be insurable for the shipowner. Accordingly, there is a conflict of interests. Any provision of limitation of liability has to strive for a balance between them.

The shipowner's right to limit his liability is admitted in the Bunkers Convention. By virtue of Articles 6 and 7(1) of the Bunkers Convention, the insurability of liability might be guaranteed, since the amount of liability required to be insured will, in all cases, not exceed "an amount calculated in accordance with the Convention on Limitation of Liability for Maritime Claims, 1976, as amended."[129] However, the amount of compensation available for victims is not certain, but depends on "any applicable national or international regime."[130]

The discussion of unlimited liability in this chapter is aimed at providing an insight to make available a possibly high level of compensation for bunker-oil pollution victims. However, the idea of "unlimited liability" was of no interest to the drafters of the Bunkers Convention. This was due to the consideration that the insurability of marine risks often depends to a great extent upon the availability of limitation of shipowners' liability.[131] What is more, if the shipowners do not or cannot take out insurance or other financial security, eligible claimants may have a problem in obtaining compensation.

---

129 The Bunkers Convention, Art.7(1).
130 The Bunkers Convention, Art.6.
131 Chen, Xia, *supra*, note 61, p.xvii.

Chapter 6: Direct Action against the Insurer and its Limited Effect

A. Introduction

Under a contract of insurance, the insurer is liable for reimbursing the assured. Except the assured who may assert the original interest, one person may avail himself of the benefits of the insurance contract when he is: (1) a person asserting the derivative interest by way of assignment; and (2) a third party for whose benefit the policy was originally undertaken, which is normally the case in liability insurance, provided that a right of direct action is granted.[1] The persons in these two circumstances may claim directly on the insurance policy, although they are not the parties to the insurance policy. In recent years, modern insurance law has increasingly given third parties who sustained damage a right to proceed directly against the liability insurer. National law often gives a third party the right to claim directly against the insurer in the case of insolvency of the insured.

As an integral part of the compulsory insurance regime under the Bunkers Convention, claims for compensation for pollution damage are entitled to be brought directly by third parties against the insurer or other party named in the certificate as the guarantor.[2] Although the Bunkers Convention does not give a definition of "direct action", it contains a detailed provision, which provides:

> "Any claim for compensation for pollution damage may be brought directly against the insurer or other person providing financial security for the registered owner's liability for pollution damage. In such a case the defendant may invoke the defences (other than bankruptcy or winding up of the shipowner) which the shipowner would have been entitled to invoke, including limitation pursuant to Article 6. Furthermore, even if the

---

1 Arnould, Joseph/Mustill, Michael J., *Law of Marine Insurance and Average* (1981), 16th ed., Volume 1, p. 155-156: "Questions have been raised as to the parties who may avail themselves of these very broad and comprehensive terms. In the first place it is clear they must be persons who may lawfully be insured. In the next place they must be persons who, at some time during the risk, have an insurable interest in the property, either as the persons originally insured or as their assignee. Beyond this, it must be shown that the person affecting the insurance either intended it for their benefit, or at all events did not intend it exclusively for the benefit of others having a conflicting or inconsistent interest, but mean it to apply generally, so as to cover the interests of those who should ultimately appear concerned..."

2 The Bunkers Convention, Art.7(10).

shipowner is not entitled to limitation of liability according to Article 6, the defendant may limit liability to an amount equal to the amount of the insurance or other financial security required to be maintained in accordance with paragraph 1. Moreover, the defendant may invoke the defence that the pollution damage resulted from the wilful misconduct of the shipowner, but the defendant shall not invoke any other defence which the defendant shall not invoke any other defence which the defendant might have been entitled to invoke in proceedings brought by the shipowner against the defendant. The defendant shall in any event have the right to require the shipowner to be joined in the proceedings."[3]

Apparently, this paragraph provides a right of direct action at length. It not only affirms the right to bring an action directly against the insurer or financial guarantor, but also entitles the defendant to invoke the defences within the limited scope. Moreover, the right of direct action is available to victims in any case once the shipowner has purchased liability insurance.[4] The basic idea of this right is to make the compensation guaranteed by liability insurance directly available to the ones who have suffered pollution damage in a bunker-oil spill. It is more convenient and secure to claim against the insurer than to pursue the shipowner. Therefore, this provision of direct action clearly endows the claimant with a very advantageous negotiation position.

Differences exist between a direct action right under the Bunkers Convention and a direct action right under national laws. This issue deserves a comparison in this chapter. In addition, the scope of claims, which mainly includes the examination of "pollution damage" arising from bunker oil-spill incidents, will be discussed.

B. Rights of a third party to claim on the insurance policy

I. Assignment

It is an established insurance principle that assignment transfers the benefits of the insurance policy to the third party. Any assignment must be of equity or of the appropriate statute.[5] A distinction is normally made between: (a) assignment of the

---

3 The Bunkers Convention, Art.7 (10).

4 By virtue of Art.7 (10) of the Bunkers Convention, the benefits of "direct action" are available to the victims whether shipowners' insurance is compulsorily taken out or not.

5 Clarke, Malcolm A., *The Law of Insurance Contract* (2002), p. 225. For instance, in England, the assignment of marine policies is now governed by the Marine Insurance Act 1906, sections 50 and 51: "50. (1) A marine policy is assignable unless it contains terms expressly prohibiting assignment. It may be assigned either before or after loss. (2) Where a marine policy has been assigned so as to pass the beneficial interest in such policy, the assignee of the policy is entitled to sue thereon in his own name; and the defendant is entitled to make any defence arising out of the contract which he would have been entitled to make if the action had

entire insurance contract, (b) assignment of the right to recover insurance money under the contract, and (c) the conclusion of a new contract between the insurer and the "assignee". The consequences of these types of assignments are different.[6] In marine insurance law, "assignment" includes the assignment of the subject-matter insured, the benefit of the contract and the policy itself.[7]

The rights of the assignee, as the third party to the original insurance contract, are thus limited by the mode of the assignment. For instance, the effect of the assignment of the subject-matter insured and the assignment of the whole insurance policy will be different. The assignment of interests in the subject-matter neither substitute the assignee for the original assured, nor does it enable him to enforce in his own name. Under the assignment of interest, the mere fact of an assured assigning or parting, by sale or otherwise, with his interest in the subject-matter of the insurance does not by itself, in the absence either of an express or of an implied agreement, have the effect of also assigning his rights under the policy.[8] In comparison, on the completion of the assignment of a policy, the rights and duties of the original assured devolve on the assignee, and the assignee becomes the assured and may accordingly enforce in his own name to all intents and purposes. The assignment of the policy is usually effected by endorsement or in other customary manners.[9] In any circumstance, if the insurance cover is assignable, the right of assignment will be limited by the manner of assignment. An assignee of a policy can only avail himself of the insurance to the extent to which the assignor has agreed to assign his rights to him,[10] and the assignee may accordingly sue against the insurer as if he were the original assured under the insurance contract.

---

been brought in the name of the person by or on behalf of whom the policy was effected. (3) A marine policy may be assigned by indorsement thereon or in other customary manner." "51.Where the assured has parted with or lost his interest in the subject-matter insured, and has not, before or at the time of so doing, expressly or impliedly agreed to assign the policy, any subsequent assignment of the policy is inoperative: provided that nothing in this section affects the assignment of a policy after loss."

6 Clarke, Malcolm A., *ibid.*, p. 226: "A distinction is made between (a) assignment of the (entire) contract of insurance, and (b) assignment of the right to recover insurance money under the contract. (a) must also be distinguished from (c), the conclusion of a new contract between insurer and "assignee". It may be difficult to distinguish (a) from (c), but it can be said that the consent of the insurer to the assignment is not of itself enough to indicate a new contract. In (a) and (c) the assignee becomes the insured under the contract, whether it be the old contract or a new contract, and the assignor drops out, a change described as a innovation, while in (b) the assignor remains the insured. This difference produces differences in the consequences of assignment. In any case, any contract which is normally assignable may be made expressly non-assignable."

7 Bennett, Howard, *The Law of Marine Insurance* (1996), p. 332.

8 Marine Insurance Act 1906, Art. 15.

9 Marine Insurance Act 1906, section 50(3): "A marine policy may be assigned by indorsement thereon or in other customary manner."

10 Arnould, Joseph/Mustill, Michael J., *supra*, note 1, p. 174.

The insurance policy can be assigned by the person in whose name the insurance has been effected to another person who may become interested therein afterwards. As far as different types of marine insurance are concerned, it is usual practice for the hull or freight insurance for a fixed time period to prohibit assignment unless specifically agreed, since the risks will be materially affected by a change of ownership of the vessel. Cargo insurance is generally left freely assignable, since once the goods have commenced transit, they normally cease to be under the close control of the cargo-owner, and broadly speaking, a change in ownership of the goods does not materially alter the risk of loss or damage in transit.[11]

Under the law of liability insurance, it is possible to assign the right to indemnity in respect of the named insured's liability or have the assignee substituted as the insured under the policy.[12] However, P&I Clubs in their Club Rules, expressly prohibit assignment. As a consequence of the event that a member of the Club has assigned the whole or any part of his interests in the insured vessel, he shall cease to be insured by the Club and the period of insurance shall terminate in respect of any vessel entered by him unless the managers of the Club have agreed in writing to continue the entry of the insured vessel, on such terms and conditions as the managers in their absolute discretion may determine.[13]

II. Direct-action statutes

If the damage is covered by the insurance policy, pollution victims certainly wish to claim directly against the liability insurer for compensation since the liability insurer is often in a more secure financial position than the shipowner. It can happen that the assured may not be in a proper financial situation before he is indemnified. Moreover, at common law, if the assured is insolvent, the money paid by his insurers after the commencement of the bankruptcy or liquidation to indemnify the insured will be added to the general assets of the insured. In such a circumstance, the victims will only have the right to bring claims in the bankruptcy or liquidation proceedings as ordinary creditors along with all other ordinary creditors of the insured, since they have no privileged right in respect of the contract between the insured and the insurer.

---

[11] Lambeth, R J., *Templeman on Marine Insurance: Its Principles and Practice* (1986), 6th Edition, pp. 75-76.

[12] Derrington, D.K, *The Law of Liability Insurance* (1990), pp. 42-43, any footnote omitted.

[13] Hazelwood, Steven J., *P&I Clubs Law and Practice* (2000), pp. 375-376: "...if he (shipowner) parts with or assigns the whole or any part of his interest in the insured vessel whether by Bill of Sale or other formal document or he otherwise ceases to have an interest in the insured vessel or if he parts with or transfers the entire control or possession of the insured vessel whether by demise charter or otherwise, unless the managers in writing have agreed to continue the entry of the insured vessel, on such terms and conditions as the managers in their absolute discretion may determine." Any footnote omitted.

In response to this disadvantageous situation of third parties, regulations concerning the right of direct action against the insurer have been adopted in various countries. The following will briefly illustrate the direct-action legislation in England and in the United States as examples.

1. English law

Common law in England does not provide for direct action. The Third Parties (Rights Against Insurers) Act 1930 (thereafter the 1930 Act) enacted by Parliament is deemed to be an act which confers on third parties rights against insurers of third-party risks in the event of the insured becoming insolvent, and in certain other events.[14] The 1930 Act confers upon third parties a "statutory subrogation". It provides the victims of an insolvent insured party with an effective right of recovery. The 1930 Act marked the starting point of a change in philosophy concerning liability insurance,[15] since it supports the view that the insurance should also be for the benefits of the injured third party rather than only for the assured.

One precondition for the operation of the 1930 Act is that the liability of the assured to the third party must be a legal liability.[16] Other aspects in relation to this right are furthermore laid down in this Act. By virtue of Article 1 of the 1930 Act, any contract made after the commencement of this Act in respect of any liability of the assured to third parties and purporting, whether directly or indirectly, to avoid the contract or to alter the rights of the parties under the policy, in the event of the assured's insolvency, shall be of no effect.[17] Meanwhile, it is a duty of the assured to provide the third-party claimants with any necessary information to enable the third parties to ascertain whether they are entitled to the rights under the Act.[18] In addition, the insurer shall be under the same liability to the third party as he would have been under to the assured except in two circumstances: first, if the liability of the insurer to the assured exceeds the liability of the assured to the third party, the assured retains the right to claim for the excess; secondly, if the liability of the insurer to the assured is less than the liability of the assured to the third party, the third party is entitled to claim against the assured for the balance.[19]

The Act, however, is perceived as having various deficiencies, which in practice undermine its purpose. After all, "...Fundamental to the 1930 Act is the principle that it is designed to transfer the assured's rights to the third party and not to create new rights or improve existing rights."[20] In certain cases, a third party

---

[14] Long title of the Third Parties (Rights against Insurers) Act 1930, reprinted in: Rose, F.D., *Marine Insurance: Law and Practice* (2004), Appendices, p. 631.

[15] Hazelwood, Steven J., *supra*, note 13, p. 309.

[16] *Ibid.*, p. 310.

[17] The 1930 Act, Art. 1(3).

[18] The 1930 Act, Art. 2.

[19] The 1930 Act, Art. 1(4).

[20] Bennett, Howard, *supra*, note 7, p. 340.

can only bring a claim against the insurer when the assured's own liability has been established or quantified.[21] In such circumstances, the claimant must thus carry through a separate action or arbitration against the assured before he can commence proceedings against the insurer under the Act.[22] Apparently, the situation might be very complicated. At the same time, the third party cannot recover his claim in the event of the assured's breach of the warranty, etc. The reason is that the third party shall not be in a better position than the assured.

The 1930 Act applies to all liability insurance.[23] The *Allobrogia* case showed that the 1930 Act could be applicable to the P&I insurance arrangement.[24] Accordingly, once a direct action right under the 1930 Act is established, a third party claiming under the Act has the right to request the Clubs to handle the claim in relation to the insured risks in the same manner as they would do if the claim were being made directly by their member. The limited effect of the statutory direct-action right against the P&I Clubs will be examined in more detail elsewhere in this chapter.[25]

---

21 *Post Office v. Norwich Union* [1967] 2 Q.B. 363, *Bradley v. Eagle Star* [1989] 1 A.C. 957.

22 Arnould, Joseph/Mustill, Michael J., *Arnould's Law of Marine Insurance and Average* (1981), 16th ed., Volume 2, pp. 1135-1136.

23 The 1930 Act, Art.1.

24 In *Re Allobrogia Steamship Corporation (The "Allobrogia")* [1979] 1 Lloyd's Rep. 190, at 194, it was held by Mr. Justice Slade: "...He referred me to footnote 14 at p. 935 of MacGillivray's and Parkington's Insurance Law (6th ed.) which reads: Difficulty sometimes arises in deciding whether a particular agreement is a contract of insurance. A mutual indemnity society which derived its funds from the contributions of its members was held to be an insurer in Wooding v. Monmouthshire and South Wales Mutual Indemnity Society Ltd., [1939]4 All E.R.570, so that an agreement to indemnify one of its members could be held to be a contract of insurance. But a person may belong to a society (such as a P. & I. Club) whose rules do not entitle him to an indemnity but only to contributions from other members towards his loss. Since the essence of a contract of insurance is that the insured should be entitled to an indemnity; it seems that in such a case there cannot be a contract of insurance. The 1930 Act contains no definition of a 'contract of insurance' but, without purporting finally to decide the point for the purpose of any subsequent proceedings, because I regard this as unnecessary for my present decision, I feel little doubt that, whatever may be the general position of Protecting and Indemnity Clubs, the relevant contracts between this particular association and its members are 'contracts of insurance' within ordinary legal terminology and within the meaning of the 1930 Act." So the judgment in this case made it clear, at least in one aspect, that the 1930 Act applies to P&I insurance. See also Chapter 2, Section C.I.4.c).

25 Section C of this chapter.

2. *United States legislation*

In the U.S., there are no federal statutory provisions concerning direct actions.[26] However, each state of the United States can enact its statute and create a right of direct action. In effect, a few states have direct-action statutes that provide victims with recourse against the assured's underwriter.[27] The benefits available to the third parties thus vary from state to state.

The state of Louisiana has established substantial statutes of the direct-action rights for third parties. It permits the third-party claimants to join the liability insurer, in its capacity as an insurer, as a defendant in the initial determining liability.[28] A direct-action claim may be brought against the insurer in Louisiana when: (a) the accident or injury occurred in Louisiana; (b) the insurance contract was issued in Louisiana; or (c) the policy was delivered in Louisiana.[29] The third party is entitled to claim without a requisite that the assured must be insolvent or at least that a judgment must be obtained against such an assured party in advance. Puerto Rico has a statute similar to the one in Louisiana. It gives the injured party the right to sue the insurer without having first obtained the judgement against the assured.[30] Puerto Rico law expressly permits direct-action suits against ocean marine insurers, including marine protection and indemnity insurers.[31]

By contrast, the right of direct action in some other states is often qualified or strictly limited. For instance, in Alabama, the attainment of a judgment against the insured is a pre-condition for bringing an action directly against the insurer. Under California law, a right of direct action is available against a vessel operator's P&I insurance by a person who has obtained a judgment against the assured. In Florida, the insurer can be sued directly under the judicial interpretation of a procedural rule that "any person may be made a defendant who has or claims an interest adverse to the plaintiff", but a third party cannot maintain a direct action against a marine liability insurer for actions accruing after October 1, 1982. New

26 Dougherty, Daniel J., 'The Impact of a Member's Insolvency or Bankruptcy on a Protection & Indemnity Club', 59 *Tul. L. Rev. 1466*, p. 1478, any footnote omitted.

27 See *ibid.*, any footnote omitted.

28 Holmes, Eric M. (general ed.), *Holmes' Appleman on Insurance* (2000), 2nd ed., Volume Fifteen, Chapters 111.1-115.3, pp. 48, 49: "The Louisiana direct-action statute was found to be substantive for conflict of law purposes and allowed the claimant to directly sue the insurer without joining the insured in the action." Footnote omitted. See also Kierr, Raymond H, 'The Effect of Direct Action Statutes on P&I Insurance, on Various Other Insurances of Maritime Liabilities, and on Limitation on Shipowners' Liability', 43 *Tul. L. Rev.*, pp. 638-672, pp. 652-657, which gives detailed information about the Louisianan Direct Action Statute and the Puerto Rican Direct Action Statute.

29 Johnson, H. Alston, 'The Louisiana Direct Action Statute', 43 *La.L.Rev.*1455, at 1478.

30 Holmes, Eric M. (general ed.), *supra*, note 28, pp. 55, 56.

31 *Reifer-Mapp v. 7 Maris, Inc.*, 1994 AMC 1215, reported in full at 830 F. Supp. 72.

York law permits a direct action by the assured against the insurer; however, this statute was made expressly inapplicable to marine insurance contracts.[32]

It is important to note that there is usually a "no-action" clause in liability insurance policy in the U.S. Early liability policy customarily contained a "no-action" clause, which provided:

> "No action shall lie against the corporation [insurer] to recover for any loss under this policy, unless it shall be brought by the assured for loss actually sustained and paid by the assured in money in satisfaction of a judgment after the trial of the issue."[33]

This type of clause is valid and will be upheld by the courts.[34] A "no-action" clause can effectively deprive injured claimants of any right under the policy, i.e. no action can be brought to recover a loss unless such an action is brought by the assured to recover for a loss actually paid in satisfaction of a final judgment or by agreement between the assured, the claimant and the insurer. "It followed, rigidly but logically, that if the insured were insolvent and perhaps declared bankruptcy (and thus never paid any amount on the judgment), no 'loss' was suffered and no amount was collectible by the victims from the insurer."[35] The insurance policy is a contract of pure indemnity with the purpose of protecting the assured's assets.[36]

Due to the fact that a direct-action right is individually permitted by the statute of a state, it is possible that where the direct-action statute applies, the "no-action" clause can be avoided.[37]

C. The limited effect of direct action under P&I insurance

Where a statutory direct-action right exists, the right of the assured to claim against his insurer is transferred to and vested in the third party to whom the liability was incurred only upon the insolvency of the assured.[38] Moreover, the effects of it under P&I insurance are rather limited.[39] P&I Clubs often try to avoid their exposure to any direct action from third-party claimants.

---

32 Maginnis, Michael J./Cot, Jose R., 'Direct Action Statutes and P&I Insurance', available at: <http://www.mlaus.org/article.ihtml?id=576&folder=90> (visited 22 February 2006).

33 Holmes, Eric M. (general ed.), *supra*, note 28, p. 35.

34 See *ibid.*, p. 34.

35 Johnson, H. Alston, *supra*, note 29, at.1457.

36 More details read Holmes, Eric M. (general ed.), *supra*, note 28, pp. 29-44.

37 *Ibid.*, p. 49.

38 For instance, in the 1930 Act, Art.1(1).

39 It has been suggested that: "although the direct action has been granted based on the general policy of protection of the weaker (third) party and the legal rules confirm this right explicitly, there seems to e a more important legal principle to protect the economic interest of the P&I Clubs." See Fossion, Gregory, 'An Eternal

I. Coverage and exclusions of P&I insurance

A shipowner obtains insurance of specified coverage from the P&I Club.[40] Therefore, the Club does not indemnify any liabilities which cannot fall within the insurance coverage agreement between the member and the Club. Additionally, it is a general principle that the Club does not extend coverage beyond the person who is named as a member and registered as a member of the Club.[41]

Except for a specific list of general insurance coverage,[42] there are specific exceptions and limitations to Club cover. The Club excludes liabilities suffered in the adventure if it is illegal, hazardous or improper. The definition of "improper" is a matter within the discretion of the directors of the Club.[43] P&I Clubs also exclude cover in respect of: (a) liability for damage to the entered vessel, loss of hire, equipment, etc.; (b) liabilities, costs and expenses of salvage vessels, drilling vessels, dredgers and others; (c) liabilities, costs and expenses arising out of or consequent upon war risks; (d) liabilities for certain nuclear risks and for the carriage of live animals; (e) liabilities due to the wilful misconduct of the member; and (f) liabilities for any loss attributed to unseaworthiness. Moreover, unless otherwise agreed, the Club's liability shall in no event exceed what would be imposed on the member by ordinary law in the absence of a contract. The P&I Club does not insure risks already insured under other insurance.[44] At the same time, the insurance coverage afforded by many Clubs is, to some extent, discretionary.

II. P&I insurance is one of indemnity

Indemnity is an established principle in marine insurance.[45] "Indemnity" denotes a shifting of liability for the loss from a person who is legally liable to another

Triangle at Sea: Loss of Insurance Cover under a Direct Action in Marine Liability Insurance', at <http://www.law.kuleuven.ac.be/jura/39n2/fossion.htm> (visited 10 October 2005).

40 Hazelwood, Steven J., *supra*, note 13, p. 113, the coverage takes great account of the individual characteristics and requirements of a member.

41 *Ibid.*, p. 84.

42 See Chapter 2, Section C.I.4.b).

43 Hazelwood, Steven J., *supra*, note 13, p. 242: "...As it may be difficult to judge what adventure are 'improper' or 'imprudent' and to avoid an unfavorable *ex post facto* definition a member may be well advised to seek the consent of the committee before engaging in an adventure which is likely to involve any unusual hazard."

44 The general reference is from *ibid.*, Chapter 9: "Exceptions and Limitation to Club Cover".

45 The indemnity principle may not be so absolute to all types of insurance. As observed in Holmes, Eric M. (general ed.), *supra*, note 28, p. 4: "The most accurate conceptualization of insurance arrangements is to observe, first, that neither life insurance nor any other form of insurance is invariably a pure indemnity contract;

person who, in the absence of insurance, is not. Under the indemnity principle, the insurer usually contracts to indemnify the assured for what he may actually lose by the occurrence of the event from which the insurer's liability is to arise. However, as it has been observed, the indemnity principle "does not always require that the amount of an insurance payment must be equal to the total actual loss"[46] and "...Insurance dispensing only partial reimbursement does not affront the indemnity principle".[47]

In most jurisdictions, P&I Clubs claim that P&I insurance is one of indemnity. The difference between liability insurance and indemnity insurance in the context of P&I insurance has been mostly discussed in the United States, which has concluded that a marine P&I policy constitutes indemnity, not liability. For instance, in *Cucurillo v. American Steamship Owners Mutual Protection and Indemnity Association, Inc*,[48] it was held that:

> "The policy (P&I policy) in this instance is not a liability policy, but an indemnification policy. As stated in 11 Couch on Insurance (2d ed.), section 44:4, at page 522: 'Thus in substance the distinction between an indemnity and a liability policy is that payment by the insured is necessary under the indemnity but not under the liability contract.'"[49]

P&I insurance insures the shipowner's liability to third parties. It is a condition that the assured shall be liable to the third party and the liability is of a kind that is covered by the insurance policy. However, the incurred liability of the assured to third parties does not determine the enforceability of the policy. The third party does not generally have the right to claim against P&I Clubs. The duty of the P&I insurer to indemnify the assured does not arise until the assured has paid damages to the third party.[50] This is more commonly referred to as the "pay to be paid" rule in the Clubs.

---

second, that all forms of insurance are subject to the influence of the principle of indemnity; and third, that the influence of the indemnity principle is less pervasive in some forms of insurance, such as life insurance, than in other forms of insurance, such as property insurance. In other words, although the characterization of insurance as an indemnity contract is useful as a statement of a tendency or as a generalization, it is not always a reliable guide when answers are sought to specific problems of insurance law."

46 Holmes, Eric M. (general ed.), *ibid.*, p. 3.

47 *Ibid.*

48 A.M.C 1969 III, p. 2334-2336.

49 A.M.C 1969 III, p. 2335. There are also cases in American: *Ali Galeb Ahmed, et al. v American Steamship Owners Mutual Protection and Indemnity Association Inc.et al.* [1978] *A.M.C.586; Robert Allen Willer v. Twin City Barge & Towing Company* [1978] A.M.C. 2008.

50 Hazelwood, Steven J., *supra*, note 13, p. 141.

III. "Pay to be paid" rule

The "pay to be paid" rule of the Club provides that the assured is not entitled to be paid by the Club until the assured himself has made payment to the third party. The typical clause to this effect is as follows:

> "Unless the committee in its discretion otherwise decides it is a condition precedent of a member's right to recover from the club in respect of any liabilities, costs or expenses that he shall first have discharged or paid the same."[51]

It is in the interests of the third party that he can recover from the Club directly if the assured is in an uncertain financial situation. However, even if the claimant may be entitled to bring an action directly against the Club,[52] these claims may be defeated by the "pay to be paid" provision under P&I insurance. In other words, if the "pay to be paid" rule or a similar clause is written in P&I insurance between the member and his Club, the statute such as the 1930 Act is inoperative. In England, the *Fanti* and *the Padre Island* cases highlighted the relationship between the principle of the "pay to be paid" rule and the 1930 Act. In both cases, the third party had a claim against the insolvent shipowner whose ship had been entered in the P&I Club. The central question was whether the 1930 Act conferred upon the third party an effective right to proceed directly against the Club for the loss or damage suffered by him despite the presence of a condition of prior payment under the Club's rules. The House of Lords ruled in favour of the P&I Clubs. The decisions were on the basis that the "pay to be paid" proviso was a term of the contract of insurance which had not been adhered to. It would not, therefore, be reasonable to confer a much more favourable right on the third party. As Lord Goff of Chieveley in the said cases stated:

> "... I start from the position that what is transferred to and vested in the third party is the member's right against the club. That right is, at best, a contingent right to indemnity, the right being expressed to be conditional upon the member having in fact paid the relevant claim or expense. If that condition is not fulfilled, the member had no present right to indemnity, and the statutory transfer of his right to a third party cannot put the third party in any better position than the member. It is as simple as that."[53]

It has been observed that this decision "seriously curtails the usefulness of the Act to third-party claimants seeking to recover claims directly from P&I Clubs whose rules are governed by English law."[54] As a matter of fact, a similar situation exists in other countries. For instance, in the United States, the P&I insurer conditions its payment to the assured on the assured first being found legally liable and actually having paid that liability before indemnifying the assured. The insurer is not liable

---

51 *Ibid.*, p. 351.

52 For instance, where a direct action statute applies.

53 *The Fanti and The Padre Island* [1990] 2 Lloyd's Rep 191, HL, at 199, per Lord Goff of Chieveley.

54 DelaRue, Colin M./Anderson, Charles B., *Shipping and the Environment* (1998), p. 721, any footnote omitted.

to indemnify the assured unless and until the assured has actually paid its own liability.[55]

IV. Other defences of a Club against the claim from a third party

If the third party may be allowed to claim directly against the P&I Club in any case as prescribed in the direct-action statute, he is not entitled to have more rights than the insured would have. The defences that a P&I Club can invoke against the third party in the first place are those that are available to the assured. Additionally, a P&I Club is also entitled to invoke other defences, which include the exclusion of insurance coverage mentioned above[56] and others, for instance, non-payment of premium, wilful misconduct of the assured and so on. "Wilful misconduct" of the shipowner deserves a detailed examination elsewhere in this chapter,[57] while the defence based on non-payment of premium will be examined below.

The insurance premium is paid to the insurer by the assured and therefore the insurer promises to indemnify the assured against the loss caused by the perils insured against. In principle, if the assured fails to pay the premium, it will constitute a fundamental breach of contract and result in loss of insurance cover for the assured. The so-called "call" under the Club Rules of the P&I Club is similar to the premium under the ordinary insurance contract.[58] It is a condition precedent to a member's right of recovery that his calls are fully paid to the Club. Most Clubs, however, also have rules providing for set-offs in respect of unpaid calls against a member's indemnity. Hence, if such a right of set-off is expressly stated, there is no reason why such a right should not also be available to third-party claimants.[59] In such a circumstance, the failure of the assured to pay calls is not a valid reason for a Club to refuse to pay to the third party if other conditions for the direct-action right are satisfied.

D. Direct-action right under the Bunkers Convention

The direct-action right under the Bunkers Convention is more than a right to enforce a claim directly as the means of avoiding procedural complexity and expenses. Due to the complexity involved in oil-spill incidents, it would be necessary to provide innocent victims with a more certain right to claim directly against the insurer. Therefore, compared with other direct-action statutes, the Bunkers Convention provides pollution victims with a more favourable position.[60]

---

55 See Section B.II.2, "no-action" clause of this chapter.
56 See Section C.I of this chapter.
57 See Section D.II of this chapter.
58 See Chapter 2, Section C.I.4.c).
59 Hazelwood, Steven J., *supra*, note 13, p. 315, any footnote omitted.
60 The Bunkers Convention, Art.7(10).

The Convention affirms its position in two aspects: first, "any claim" for compensation for pollution damage may be brought directly against the insurer or guarantor. The phrase "any claim" is broad enough even to include claims from the shipowner, since "pollution damage" includes the risks and costs incident to any preventive measures taken by the shipowner.[61] Nevertheless, the shipowner does not need such a direct-action right since he is a party to the insurance contract. Additionally, claims can be brought against any person providing financial security for the registered owner's liability for pollution damage once the registered owner has it in place, whether voluntarily or compulsorily.[62]

Secondly, the insurer or guarantor has a very limited scope of defences. Pursuant to Article 7(10) of the Bunkers Convention, the claimant is liable to be defeated by any defence invoked by the insurer which the shipowner would have been entitled to invoke, save only that the Convention nullifies clauses which purport to relieve the insurer from liability upon the shipowner's insolvency.[63] In addition, the insurer or guarantor may invoke the defence that pollution damage resulted from the wilful misconduct of the shipowner. The following sections will analyse the defences of the insurer in detail.

At the same time, the exposure of the insurer or guarantor shall not exceed the liability limit applying to the ship under the Bunkers Convention. The insurer or guarantor can, even in the case of the shipowner not being entitled to limitation of liability, take advantage of limitation of liability. More will be discussed below. The insurer can also have the right to require the assured to join in legal proceedings brought against the insurer.[64]

If a liable person other than the registered owner is liable and insurance or other financial security has been in place, it is questionable whether claims for compensation for pollution damage can be brought directly against the insurer in this case. Article 7(10) provides that "Any claim for compensation for pollution damage may be brought directly against the insurer ... providing financial security for the registered owner's liability for pollution damage..." Only the right to claim directly against the liability insurer of the registered owner is expressly given in the Bunkers Convention.

---

61 The Bunkers Convention, Art. 1(9) (b), Art. 1(7).

62 The Bunkers Convention, Art.7(10). Since only the registered owner of the ship having a gross tonnage greater than 1,000 is required to maintain insurance or other financial security, Art.7(1).

63 Under existing insurance law, the insurer is liable only after the assured has actually paid out on a claim in certain cases, such as the "pay to be paid" rule in P&I insurance. However, this is not a defence available to the insurer or guarantor in the Bunkers Convention. This position is shown in the brackets in Article 7(10): "...In such a case the defendant may invoke the defences (other than bankruptcy or winding up of the shipowner)..."

64 The Bunkers Convention, Art. 7(10).

I. "the defences ... which the shipowner would have been entitled to invoke"

If any defences had been available to the shipowner, it is equally important for the insurer or guarantor to be able to avail himself of these defences. Accordingly, the insurer may also rely upon any exceptions which would have been applicable to the shipowner in a particular case.

By virtue of Article 3(3)[65], Article 3(4)[66], and Article 3(5)[67] of the Bunkers Convention, no or only a part of liability for pollution damage shall attach to the shipowner if the damage was caused by external factors such as an Act of God, exceptional natural phenomena, *force majeure*, or fault of the victim or of a third party. Accordingly, the liability insurer or guarantor also has the possibility to escape from or mitigate his liability by establishing that the damage was caused by those external factors.

The insurer or guarantor may invoke the defence of the right to limit liability if the shipowner is entitled to do so. In the light of Article 7(10), the guarantor's right to limit his liability is protected in two different circumstances. First, he is entitled to limit his liability pursuant to Article 6 of the Convention. Article 6 confers a right on the shipowner and his liability insurer to limit his liability under any applicable national or international law.[68] Secondly, if the shipowner is not entitled to limit his liability according to Article 6, his insurer can still limit the liability to an amount equal to the amount of insurance or other financial security required to be maintained in accordance with Article 7(1). In the former case, if the shipowner has the right to limit his liability, his liability insurer will benefit as well. In the latter, the liability insurer enjoys a better position than the shipowner. He can limit his liability although the shipowner is not entitled to limit his liability. The insurer's liability does not exceed a limitation amount contained in the Convention on Limitation of Liability for Maritime Claims, 1976, as

---

65 The Bunkers Convention, Art.3(3): "No liability for pollution damage shall attach to the shipowner if the shipowner proves that: (a) the damage resulted from an act of war, hostilities, civil war, insurrection or a natural phenomenon of an exceptional, inevitable and irresistible character; or (b) the damage was wholly caused by an act or omission done with the intent to cause damage by a third party; or (c) the damage was wholly caused by the negligence or other wrongful act of any Government or other authority responsible for the maintenance of lights or other navigational aids in the exercise of that function."

66 The Bunkers Convention, Art.3(4): "If the shipowner proves that the pollution damage resulted wholly or partially either from an act or omission done with intent to cause damage by the person who suffered the damage or from the negligence of that person, the shipowner may be exonerated wholly or partially from liability to such person."

67 The Bunkers Convention, Art.3(5): "No claim for compensation for pollution damage shall be made against the shipowner otherwise than in accordance with this Convention."

68 The Bunkers Convention, Art. 6.

amended.[69] Therefore, in the latter case, if the guarantor's exposure is limited to a lower amount than the shipowner's liability, the claimants' recovery for any excess will depend on applicable law in a particular case.

II. Meaning of the phrase "wilful misconduct of the shipowner"

The insurer or the guarantor is free from liability if pollution damage resulted from wilful misconduct of the shipowner. The burden of proving wilful misconduct and whether it was a proximate cause of pollution damage is on the liability insurer or the guarantor. The precise meaning of "wilful misconduct" is influenced by public policy consideration and therefore varies according to the context.[70] What constitutes "wilful misconduct" under the Bunkers Convention? In the absence of a definition in the Bunkers Convention, the interpretation of "wilful misconduct" may in effect depend on which national court decides on the issue.

1. Interpretation of "wilful misconduct" in relation to limitation of liability

One can find that the term "wilful misconduct" was discussed in some international conventions in relation to limitation of liability. "Wilful misconduct" was chosen to expressly describe conduct barring limitation of liability in the 1929 Warsaw Convention.[71] Article 25(1) of the said convention provided that, if the plaintiffs proved that "the damage is caused by his wilful misconduct or by such default on his part as, in accordance with the law of the Court seised of the case, is considered to be equivalent to wilful misconduct",[72] the defendant could not avail himself of the liability limits.

Furthermore, by virtue of a "default...is considered to be equivalent to wilful misconduct", the said convention had to be drawn in such a way that it would be applied by civil and common law courts.[73] Accordingly, if the term "wilful misconduct" did not exist in the law of a particular nation in which a case in relation to the Warsaw Convention was being tried, any issue relating to "wilful misconduct" in that case was to be considered under whatever principle was equivalent to "wilful misconduct" in that nation's legal system. In other words, the court to which the case was submitted had the right to measure the conduct by its own standards.

---

69 The Bunkers Convention, Art.7(1) and Art.7(10).

70 Clarke, Malcolm A., *supra*, note 5, p. 802.

71 The 1929 Warsaw Convention was agreed in 1929 between a number of countries to establish uniform legislation affecting the legal responsibility of international air carriers. It relates only to international flights.

72 The Warsaw Convention, Art. 25(1), text of United Kingdom translation, reprinted in: Goldhirsch, Lawrence B., *The Warsaw Convention Annotated: a Legal Handbook* (2000), in Appendix, pp. 245-257.

73 Goldhirsch, Lawrence B., *ibid.*, p. 152.

The concept of "wilful misconduct" exists in common law courts. In England, "wilful misconduct" has been interpreted by case laws. In *Forder v. Great Western Railway Company*[74], based on the definition of "wilful misconduct" given by Johnson J. in *Graham v. Belfast and Northern Counties Ry. Co.*[75], wilful misconduct "in such a special condition" means "misconduct to which the will is party as contradistinguished from accident, and is far beyond any negligence, even gross or culpable negligence, and involves that a person wilfully misconducts himself who knows and appreciates that it is wrong conduct on his part in the existing circumstances to do, or to fail or omit to do it, or persists in the act, failure, or omission regardless of consequences."[76] Lord Alverstone C.J. adopted an addition, which was: "...or acts with reckless carelessness, not caring what the results of his carelessness may be."[77] By American Courts, it has been observed that, wilful misconduct applies to two different types of acts or omissions, namely: "(1) an intentional act or omission to act, and (2) an act or omission to act with reckless indifference or disregard as to the probable consequences of the act or omission."[78]

According to the commentary on the Warsaw Convention, it was hard to find a term equivalent to "wilful misconduct" which was recognised in common law in civil law. The draft of Article 25 submitted to the Warsaw Conference predicated unlimited liability of the carrier on "*dol*" on his part or on the part of his servant or agent.[79] The French word "*dol*" was deemed to be equivalent to "wilful misconduct" at the time. It implied an act or omission that was done with the intent to cause damage. However, there are differences between "*dol*" and "wilful misconduct", since, at least, the latter can comprise "a reckless act or omission with the knowledge, sometimes implied, that harm will occur."[80] The position also varies when discussing whether "*faute lourde* (gross negligence)" is equivalent to "*dol*". Under Austrian, German, Swedish and Swiss law, "*faute Lourde*" is equivalent to "*dol*", but Belgian and Italian law does not permit such assimilation.[81]

Due to the confusion and controversy involved in the interpretation of the said Article 25 in the Warsaw Convention, the stipulation of "wilful misconduct" was substituted by "...an act or omission of the carrier, his servants or agents done with intent to cause damage or recklessly and with knowledge that damage would probably result..."[82] The new Article defines "wilful misconduct" in terms of a person's actions.[83]

---

74 [1905] 2 K.B.532.

75 [1901] 2 I.R.13.

76 [1905] 2 K.B.532, at 535.

77 *Ibid.*, at 536.

78 Mankiewicz, Rene H, *The Liability Regime of the International Air Carrier* (1981), p. 126.

79 Mankiewicz, Rene H, *ibid.*, p. 126.

80 Goldhirsch, Lawrence B., *supra*, note 72, p. 152.

81 Mankiewicz, Rene H, *supra*, note 78, pp. 122,123.

82 See Article XIII of the Hague Protocol, 1955, reprinted in: Goldhirsch, Lawrence B., *supra*, note 72, in Appendix, pp. 311-320.

83 Goldhirsch, Lawrence B., *ibid.*, p. 151.

As a matter of fact, "wilful misconduct" was discussed as a term to be also used to describe conduct barring limitation of liability for the 1969 CLC. It was not accepted, since it was observed that "wilful misconduct", which had a meaning in England, was not known in the law of continental Europe.[84] A different wording was suggested as more desirable. The 1969 CLC thus adopted "actual fault or privity" as conduct barring the shipowner's right to limit.[85] A lengthy provision was opted for in the 1984 Protocol to the CLC,[86] which provides that the owner shall not have a right to limit his liability if it is proved that pollution damage "resulted from his personal act or omission, committed with the intent to cause such damage, or recklessly and with knowledge that such damage would probably result".[87] Therefore, if the intention to prohibit the shipowner from limiting his liability in the case of his own wilful misconduct was retained, this lengthy provision should convey a similar meaning to "wilful misconduct". This idea is shared by some authors.[88] A similar provision is followed in the CLC Protocol 1992.

This approach is adopted by other conventions. Article 4 of the 1976 LLMC stipulates that a person liable shall not be entitled to limit his liability if "it is proved that the loss resulted from his personal act or omission, committed with the intent to cause such loss, or recklessly and with knowledge that such loss would probably result."[89] This rule is widely accepted. It may be the limitation of liability rule in a bunker-oil spill incident if the applicable limitation regime under the Bunkers Convention is the 1976 LLMC.[90] In fact, some countries which are not parties to the 1976 LLMC have also adopted similar rules in this regard. For instance, China is not a Party State to the 1976 LLMC, but it provides in Article 209 of the Maritime Code 1993 that: "A person liable shall not be entitled to limit his liability…if it is proved that the loss resulted from his act or omission done with the intent to cause such loss or recklessly and with knowledge that such loss would probably result."[91] The Chinese Court has observed that it may sometimes regard a serious fault of the vessel owner as evidence of a reckless act or omission.[92]

---

84 O.R. 1969, p. 733.

85 The 1969 CLC, Art.V(2). State parties may give different interpretations to "fault".

86 See Chapter 1, Section A.II.

87 See 1984 Protocol, Art.6, reprinted in: *Lloyd's Shipping Law Library*, *The ratification of Maritime Conventions* (2004), Vol.4, II.7.50.

88 For instance, Wu Chao, *Pollution from the Carriage of Oil by Sea: Limitation and Compensation* (1996), p. 175: "Therefore, the 1984 Protocol to the CLC replaced the test of 'actual fault or privacy' with one of 'wilful misconduct'." Footnotes omitted.

89 The 1976 LLMC, Art.4.

90 The Bunkers Convention, Art. 6.

91 The Maritime Code of the People's Republic of China 1993, Art.209, reprinted in: Li, KX/Ingram, CWM, *Maritime Law and Policy in China* (2002), p. 55.

92 Mo, John Shijian, *Shipping Law in China* (1999), p. 317.

2. *Interpretation of "wilful misconduct" in insurance law*

a) The determination of a competent court

Before wilful misconduct can be ascertained under Article 7(10) of the Bunkers Convention, the place where an action can be brought must first be considered.

The assured's conduct shall be judged according to the law governing the relationship between the insurer and the assured. In this respect, it is important to note that the Rulebooks of the P&I Clubs often provide a clause which refers disputes between a member or a co-assured and the Club to the specific jurisdiction of one court or to arbitration.[93]

Pursuant to Article 9(1) of the Convention, actions for compensation may only be brought in courts of Contracting States where pollution damage has occurred or preventive measures have been taken to prevent or minimise such damage.[94]

b) "Wilful misconduct" in P&I insurance

"Wilful misconduct" is not covered by any insurance policy, since "insurance is against fortuitous events only."[95] It is also typical that P&I Clubs in the International Group exclude insurance cover where wilful misconduct can be proved. It is a ground on which a liability insurer or guarantor will have a defence against direct action against him under the Bunkers Convention.[96] Wilful misconduct may occur at any time during the insurance period and its effect on the subject-matter of the insurance may only be unavoidable when the conduct occurs.[97]

There are different terms such as negligence, recklessness and wilful misconduct. Whether a negligent act can be wilful is questionable. The wilful nature is more than just acting negligently. However, it may be difficult to draw a line between wilful misconduct and negligence.[98] In order to deprive the assured of insurance cover based on "wilful misconduct", it is necessary to show that the assured's wilfulness was directed not only towards a particular act or omission, but also toward the "misconduct." In other words, the mental element of "wilful" should not be divorced from the consequence which amounts to "misconduct." Additionally, "misconduct" should first of all be wrongful under the contractual or other relationships at the relevant time. That is to say, in the context of a contract, one party must do or omit to do something which is aptly described as misconduct towards the other contracting party.[99] It has been observed that the most prevalent example of the use of this defence by the insurer is the case of the deliberate sinking of a ship in order to make a fraudulent claim under the insurance policy.[100]

---

93 Gauci, Gotthard, *Oil Pollution at Sea* (1997), pp. 252, 253.
94 The Bunkers Convention, Art. 9(1).
95 Merkin, Robert M. (ed.), *Colinvaux's Law of Insurance* (1997), p. 91.
96 The Bunkers Convention, Art. 7(10).
97 Clarke, Malcolm A., *supra*, note 5, p. 490.
98 Merkin, Robert M. (ed.), *supra*, note 95, p. 91.
99 *National Oilwell (UK) Ltd v Davy Offshore Ltd* [1993] 2 Lloyd's Rep. 582, p. 622, per Colman J.
100 *Ibid.*, p. 621.

The contract of insurance between the Club and its member is normally governed by the Club Rules with the modification agreed by the Club manager. The Rules in most Clubs, except a few which are in the U.S., Japan, Norway, and Sweden[101], are governed or construed in accordance with English law.[102] In English law, loss or damage attributable to wilful misconduct of the assured is not recoverable under the marine insurance policy. The stringent consequence is due to section 55(2)(a) of the Marine Insurance Act 1906, which provides: "The insurer is not liable for any loss attributable to the wilful misconduct of the assured..."[103] At the same time, this exclusion is not conditioned on the phrase of "unless the policy otherwise provides."

In England, it is often preferred in relation to the facts of the case to give an example of circumstances in which wilful misconduct might arise rather than establishing a definition of wilful misconduct.[104] In *Papadimitriou v Henderson*,[105] the characteristics of an act of wilful misconduct were discussed in the context of war risks as follows:

> "...of course, if it was a case in which the shipowner got warning that a blockade had been established at a particular port or that a ship was lying waiting at a particular point, and the shipowner deliberately sent his ship forward to that point to run the blockade, it may be that there would be, in certain cases, an inference to be drawn that he was not endeavouring to carry out the voyage, but what he was endeavouring to do was to get his ship captured, and that, of course, would be wilful misconduct."[106]

3. *"Wilful misconduct" in Article 7(10)*

By virtue of Article 7(10) of the Bunkers Convention, the insurer or guarantor is free from liability if the pollution damage resulted from wilful misconduct. In other areas of law in relation to insurance contract, the finding of fault in subordinates and agents has been traced upwards to the *alter ego* of the ship-owning company.[107] It is doubtful whether the same principle can or should apply to the oil-pollution case. In effect, the proposition in this paragraph extends only to the wilful misconduct of the shipowner. In other words, it must be the shipowner who was the author of "wilful misconduct." Any "wilful misconduct" committed by the registered owner, the bareboat charterer, the manager and the operator will preclude the shipowner from recovering from his liability insurers.

---

[101] DelaRue, Colin M./Anderson, Charles B., *supra*, note 54, p. 699, F.N. 13.

[102] *Ibid.*

[103] Marine Insurance Act 1906, s.55(2)(a).

[104] *National Oilwell (UK) Ltd. v. Davy Offshore Ltd.* [1993] 2 Lloyd's Rep. 582, at 621.

[105] [1939] 64 Ll.L.Rep. 345.

[106] *Ibid.*, p. 349, per Goddard LJ.

[107] Hazelwood, Steven J., *supra*, note 13, p. 246. As to a company's *alter ego*, see *H.L. Bolton Engineering Co.Ltd v. T.J.Graham&Sons Ltd.* [1957] 1 Q.B. 159, pp. 172 and 173, per Denning L.J.; see also *Lennard's Carrying Co.Ltd. v. Asiatic Petroleum Co. Ltd* [1915] A.C. 705.

However, the loss caused by the wilful misconduct of the shipowner's servant may not free the shipowner's insurer from his liability.

E. Scope of the claim: pollution damage

I. The uncertain nature of liability for pollution

Liability insurance relates to the nature, extent and amount of liability. It contributes to the effective operation of insurance when the frequency and severity of liability are predictable, as it has been suggested that: "...Insurance operates most comfortably with stochastic events, in which the probability of the frequency and magnitude of insured losses that will be suffered by a group of policyholders is highly predictable..."[108] However, when the liability is uncertain, it is difficult, if not impossible, to insure it.

It has been observed that some uncertainties exist in relation to environmental liability insurance. First of all, scientific uncertainty and ambiguities "make it virtually impossible to assemble a predictive database for assessing risk".[109] It may consequently limit the amount of insurance coverage available for such risks.[110] This uncertainty can be reduced if there are sufficient scientific and technical methods to determine the nature of the risk and prevent the occurrence of pollution incidents in some cases. The insurers can therefore find guidelines to measure the activities of those they insure and provide liability insurance.

In addition, the laws regulating environmental liability are changing. This makes the extent of environmental liability hardly predictable. The most obvious example is that under the U.S. Superfund,[111] which imposes retroactive strict liability on the insurer.[112] In such a circumstance, the insurer may take precautionary measures to avoid liabilities by inserting in the insurance policy a provision declaring that he will not be liable if the liabilities of the assured are susceptible to legal changes. The uncertainty is also present in other aspects of insuring liability for pollution. Comparatively, the law governing oil pollution liability on an international level is stable and specific with regard to the definition and scope of "pollution damage".

---

108 Abraham, Kenneth S., 'Environmental Liability and the Limits of Insurance', 88 *Colum. L. Rev.* (1988), pp. 942-988, at 946, 947.

109 Kunreuther, Howard, 'The Role of Insurance and Compensation in Environmental Pollution Problems', workshop papers, May 15, 1986, Center for Research on Risk and Insurance, the Wharton School, University of Pennsylvania with the contributions of the Geneva Association, pp. 1-6, at 2.

110 See *ibid.*

111 The Comprehensive Environmental Response, Compensation, and Liability Act of 1980, commonly known as Superfund.

112 Tanega, Joseph, 'Implications of Environmental Liability on the Insurance Industry', (1996) 8 *J. Envtl. L.*, pp. 115-137, at 129.

II. "Pollution damage" in tanker-oil spill incidents

A traditional construction of the concept of "pollution damage" generally includes the following damage: (1) loss of life, personal injury or damage to property or the environment; (2) loss of profit or income; and (3) costs of reinstatement or preventive measures.[113] These three types of damage may also occur in any oil-spill incident.

The 1969 CLC first defines "pollution damage" caused by an oil spill as "loss or damage caused outside the ship carrying oil by contamination resulting from the escape or discharge of oil from the ship, wherever such escape or discharge may occur, and includes the costs of preventive measures and further loss or damage caused by preventive measures."[114] The CLC Protocol 1992 contains an amended definition of "pollution damage" which was very much based on the original one. However, it added a phrase to clarify that only costs incurred for reasonable measures to restore the contaminated environment are considered as "environmental damage" within the definition of "pollution damage".[115] This can fairly avoid speculative claims.

According to the definition in the CLCs, the pollution damage which may occur in an oil-spill incident include: (1) loss or damage caused by contamination; (2) the costs of preventive measures; (3) the costs for the damage caused by preventive measures. Briefly, since loss or damage must be caused outside the ship, the right to compensation thus goes to claimants who have no connection with the vessel.[116] Since the loss or damage must be caused by contamination, loss by fire is not covered. But it has been suggested that contamination by oil following a fire would, however, be covered.[117] The contamination should result from the escape or discharge of oil from the ship and there is no restriction on whether it is an operational or accidental discharge. Preventive measures are furthermore defined as meaning "any reasonable measures taken by any person after an incident has occurred to prevent or minimise pollution damage".[118] This paragraph denotes three elements: first, preventive measures can be taken by any person, including

---

[113] Brodecki, Z., 'New Definition of Pollution Damage', *Lloyd's Mar. & Com. L. Q.* (1985), pp. 382-391, at 386. These three elements are largely reflected in the CLCs.

[114] The 1969 CLC, Art. 1(6).

[115] The CLC Protocol 1992, Art. 2(6): "...provided that compensation for impairment of the environment other than loss of profit from such impairment shall be limited to costs of reasonable measures of reinstatement actually undertaken or to be undertaken;..." Meanwhile, according to Jacobsson, Mans/Trotz, Norbert, 'The Definition of Pollution Damage in the 1984 Protocol to the 1969 Civil Liability Convention and the 1971 Fund Convention', 17 *J. Mar. L. & Com.* 467 (1986), at 487, 488: "...The new definition clearly excludes compensation based on a theoretical calculation of damage caused to the marine environment by oil without actual proof of the costs of reinstatement."

[116] Wu, Chao, *supra*, note 88, p. 46.

[117] Abecassis, David W. (ed.), *Oil Pollution from Ships* (1985), pp. 208, 209.

[118] 1969 CLC, Art. 1(7).

the shipowner. Secondly, the measures shall be reasonable and for the purpose of preventing or minimising pollution damage. Thirdly, the measures are taken after but not before the incident occurred. The definitions of "incident" are different in two CLC conventions. The CLC Protocol 1992, on the basis of the 1969 CLC, furthermore includes any occurrence or series of occurrence having the same origin which creates a great and imminent threat of causing such damage.[119] The CLCs also allow compensation to be available for further loss or damage caused by preventive measures. Hence, it has been pointed out "damage caused by the use of dispersants is covered in the Convention, irrespective of whether it was brought about by contamination, fire, explosion or other causes".[120]

It is a common approach in international civil liability conventions to have a general definition of "damage" and thus give considerable freedom of interpretation to the courts of the contracting States.[121] However, "for the functioning of this regime",[122] i.e. for the operation of the common IOPC Fund, it is important to maintain a uniform treatment of claims for oil-pollution damage internationally. Therefore, some institutions have contributed to establishing a set of uniform rules to allow victims' claims resulting from a tanker-oil spill. All standards applicable to a tanker-oil spill incident may also help to clarify the types of pollution damage under the Bunkers Convention. The survey will focus on the guidelines or policy adopted by three different institutions: the CMI, the IOPC Fund and the P&I Clubs.

1. *CMI guidelines on oil-pollution damage*

The CMI Guidelines on Oil Pollution Damage were adopted at the 35th International Conference of the CMI, which was held in Sydney from 2-8 October 1994.[123] The reasons for CMI's concern was due to the following reasons: first, its long-term commitment to the unification of maritime law; secondly, the internal defects of the international system in respect of admissibility and assessment of claims for oil pollution damage; thirdly, the warning from the OPA 90, which has intensified national divergence in dealing with oil-pollution damage.[124]

---

[119] "Incident" is defined in the conventions, for instance, 1969 CLC, Art. I(8), 1992 CLC, Art.2(4). Also see Chapter 1, Section C.III.

[120] Wu, Chao, *supra*, note 88, p. 50, any footnote omitted.

[121] Jacobsson, Mans/Trotz, Norbert, *supra*, note 115, at 481.

[122] This refers to CLCs and the Fund conventions. See *ibid.*, p. 481, 482: "It was argued that a uniform interpretation of the definition of pollution damage was essential for the functioning of this regime. This was considered particularly important in respect to the Fund Convention, as under that Convention oil receivers in one Contracting State contribute to the payment of compensation for damage sustained in other contracting States."

[123] Available at: <http://www.uctshiplaw.com/cmi/cmioil.htm> (visited 12 June 2005): "Introductory note".

[124] As stated in the Guidelines: "In the early 1990s, however, there were concerns that increasing public interest in environmental issues would lead to growing problems resulting from legal uncertainties in this field. These were highlighted in 1990

The Guidelines have the following purposes: "first, to state the extent to which claims are thought to be recoverable under the law as applied in the majority of countries, and with due account being taken also of the criteria developed by the International Oil Pollution Compensation Fund; secondly, to employ terminology whose meaning is understood and acceptable in countries with a variety of different legal traditions; and thirdly, to strike a satisfactory balance between the desire on the one hand for greater certainty as to the types of recoverable claim, and on the other the need to retain sufficient flexibility to deal on their merits with the many different types of claims which may be made in practice."[125]

The Guidelines include three parts: "General", "Economic loss" and "Preventive measures, clean-up and restoration". In the "General" part, the Guidelines emphasise that compensation may be refused or reduced if a claimant fails to take reasonable steps to avoid or mitigate any loss, damage or expense. The Guidelines give no further interpretation of "reasonable", but it uses the term "reasonable" several times and tries to draw a line between admissible and inadmissible claims for compensation. For instance, in the part on "Economic loss", compensation is declared to be payable for consequential loss and pure economic loss, although pure economic loss must be caused by the contamination and pure economic loss will be treated as caused by contamination only when a reasonable degree of proximity exists between the contamination and the loss. The cost of preventive measures is recoverable insofar as both the measures themselves and the cost thereof were reasonable in the particular circumstance.[126]

The Guidelines do not change legal rights in any way.[127] As the CMI is a private organisation, its guidelines do not intend to have any force.[128] They may only assist national courts when they are faced with the task of determining difficult issues as regards the concept of "pollution damage".

2. *The policy adopted by the IOPC Fund*[129]

The IOPC Fund is a worldwide intergovernmental organisation established under the Fund Conventions to provide supplementary compensation for tanker-oil

when the USA decided to join the international system, but to adopt its own laws in the form of the US Oil Pollution Act (OPA). The Act set out a detailed framework of compensation quite different from the concept of 'pollution damage' as defined in the Civil Liability and Fund Conventions, and the Protocol thereto. This definition is couched in only very general terms, leaving scope for uncertainty as to the types of recoverable claim. It was foreseen that divergent decisions in different national courts could seriously undermine the uniform application of the Conventions which is so important for their success."

125 In CMI Guidelines on Oil Pollution Damage, *supra*, note 123.

126 More discussion see Özçayir, Z. Oya, *Liability for Oil Pollution and Collisions* (1998), p. 255.

127 *Ibid.*

128 *Ibid.*

129 More analysis see *ibid.*, pp. 242-254.

pollution victims.[130] The policy adopted by the IOPC Fund as regards the assessment and admission of claims is very important. It has accumulated considerable experience with regard to the admissibility of claims for compensation for oil-pollution damage.

The IOPC Fund can only accept those claims which fall within the definition of pollution damage and preventive measures laid down in international civil liability conventions. According to the IOPC Claims Manual, there are general criteria applicable to all claims as follows:

> (a) any expense/loss must actually have been incurred;
> (b) any expense must relate to measures which are deemed reasonable and justifiable;
> (c) a claimant's expense/loss or damage is admissible only if and to the extent that it can be considered as caused by contamination;
> (d) there must be a link of causation between the expense/loss or damage covered by the claim and the contamination caused by the spill;
> (e) a claimant is entitled to compensation only if he has suffered a quantifiable economic loss;
> (f) a claimant has to prove the amount of his loss or damage by producing appropriate documents or other evidence.[131]

Regarding the proof of "loss or damage", a claim is admissible only to the extent that the amount of loss or damage is actually demonstrated. Some flexibility is needed as regards the requirement to present documents in some particular circumstances. All elements of proof are considered and the evidence provided must give the IOPC Fund the possibility of forming its own opinion on the amount of the loss or damage actually suffered.[132]

For any specific claim to fall within the Fund, it is necessary to assess its particular characteristics, which are sometimes quite problematic and protracted. The survey confirms that claims in respect of preventive measures (including clean-up), property damage, economic loss and environmental damage may be accepted by the Fund. The Assembly and the Executive Committee have given numerous important decisions as regards the settlement of claims. For instance, in respect of environmental damage, the IOPC Fund maintain its position that claims relating to the impairment of the environment should be accepted only if the claimant has sustained a quantifiable economic loss and that loss must be such that it can be quantified in monetary terms.[133]

---

[130] See Chapter 1, Section A.II. Also see *ibid.*, p. 242: "The aim of the Fund is to settle claims out of court to provide compensation for the claimants as soon as possible. But the claimants also have the right to take their claims to the competent national court."

[131] IOPC Claims Manual Nov. 2002, available at: <http://www.iopcfund.org/npdf/92claim.pdf> (visited 12 June 2005), at 18-19.

[132] *Ibid.*, at 18.

[133] Özçayir, Z. Oya, *supra*, note 126, p. 254.

3. *Viewpoint of the P&I Clubs*

Before the establishment of international civil liability conventions, most of the cases involving oil-pollution damage were settled out of courts by the P&I Clubs,[134] the Clubs have thus gathered great experience in admissibility of claims for pollution damage. After the ratification of the CLCs and the Fund Conventions, P&I Clubs have worked closely together with the IOPC Fund in most tanker-oil spill incidents, which include the investigation, handling and settlement of claims.[135] If the payment of compensation has been made by the member to pollution victims, the Club will indemnify it. All eligible victims can also claim directly against the Club according to international civil liability conventions.

In general, there is no specific definition of pollution damage in Club cover. The Club provides insurance coverage for pollution which mainly includes: (1) legal damages and compensation payable by the members in respect of pollution incidents; (2) prevention and clean-up expenses; (3) certain expenses incurred by the owners in order to comply with orders and directions given by the authority of the State affected by the pollution incidents; (4) "safety-net" payment to the salvors; (5) contractual liabilities involved by the member which might be subject to the payment of an additional premium; (6) fines.[136] The Club will compensate victims who suffer pollution damage as defined in international civil liability conventions. The criteria may be identical to what has been applied by the IOPC Fund.[137]

III. Pollution damage under the Bunkers Convention

Following the format in the CLCs, the Bunkers Convention provides nearly the same definition of "pollution damage" as in the 1992 Protocol to CLC.[138] This definition is in general terms, which leaves a great deal to the interpretations of pollution damage at the domestic level. It would be more desirable to have a more precise definition. However, this definition may be acceptable, since a fund such

[134] Jacobsson, Mans/Trotz, Norbert, *supra*, note 115, at 470.

[135] See Chapter 2, Section C.II.1.d).bb).

[136] Hazelwood, Steven J., *supra*, note 13, pp. 218-228.

[137] Except some claims which are covered by the IOPC Fund but not covered by the liability conventions; meanwhile, as observed by Özçayir, Z. Oya, *supra*, note 126, p. 248, "The P&I clubs support the view of the IOPC Fund that a uniform interpretation of the definition of 'pollution damage' is essential for the proper functioning of the regime of compensation established under the CLC and FC. The clubs agree...But they urge a cautious and restrictive approach to the criteria for 'preventive measures' when considering economic loss. The clubs support the decisions of the Executive Committee in relation to environmental damage but emphasize that an environmental damage claim is not payable when the claim is not quantifiable by reference to an actual economic loss."

[138] The Bunkers Convention, Art.1(9).

as the IOPC Fund does not exist for bunker-oil pollution claims and it will not have such a great effect even if any courts give it an "extensive interpretation."[139]

Due to the similarities of the definitions of "pollution damage" given by the CLCs and the Bunkers Convention, for the assessment of pollution damage caused by a bunker-oil spill, past experience and academic research results regarding pollution damage in tanker-oil pollution incidents may be extremely helpful and can be used as a reference. Nevertheless, there is a fundamental difference between the definition of "pollution damage" under the Bunkers Convention and the earlier civil liability conventions, since the Bunkers Convention only covers the damage caused by bunker oil, which means "any hydrocarbon mineral oil including lubricating oil, used or intended to be used for the operation or propulsion of the ship, and any residues of such oil."[140] Accordingly, the pollution damage must involve bunker oil.

F. Recourse action

The compensation under the Bunkers Convention can basically involve three parties: the liability insurer, the assured and the third parties. It is relevant to clarify certain recourse action rights between them.

Generally speaking, the Bunkers Convention, in its Article 3(6), states that nothing in this Convention shall prejudice any right of recourse of the shipowner which exists independently of this Convention. For instance, the owner may pursue the owner of a ship which collided with his ship intentionally or by negligence and caused it to spill oil.[141] The basis of this right can be in tort, in contract or follow other principles. Meanwhile, should a liability insurer be compelled to pay oil-spill claimants due to the recognised right of direct action, the question is whether the insurer is entitled to an indemnity from the assured where a defence would have been available against the claim under the insurance policy. This question is not addressed in the Bunkers Convention; it thus has to be decided by relevant terms of insurance contract or by the law governing the contract.[142] The case can be so complicated that necessary reference may have to be made to certain principles of private international law.

In addition, the Convention does not indicate whether the shipowner's right of recourse can be assigned to or be subrogated by the liability insurer if the victim claims directly against the insurer. This has to be decided by the applicable

139 Jacobsson, Mans/Trotz, Norbert, *supra*, note 115, at 482.

140 The Bunkers Convention, Art.1(5).

141 Wu, Chao, *supra*, note 88, p. 56: "if, for example, the pollution incident is due to faulty construction of the vessel, the shipowner can bring an action under ordinary law against the shipbuilder. Similarly, if pollution were due to the negligence of anther colliding ship which is not a 'CLC-ship', the shipowner is entitled to sue the owner of the colliding ship under the principles of ordinary maritime law."

142 Similar question exists in the CLCs, see DeLaRue, Colin M./Anderson, Charles B., *supra*, note 54, p. 115.

insurance law. Subrogation is different from assignment. As previously discussed, assignment means that a person who has no original interest in a policy may have a derivative interest through assignment.[143] Aside from assignment, there is a doctrine of subrogation in insurance law. Where the insurer pays the assured, he is thereby subrogated equitably or legally[144] to all rights and remedies of the assured in respect of the subject-matter as from the time when the casualty caused the loss. The insurer is entitled to the benefits of such rights and all remedies as the assured would himself be able to enjoy. Meanwhile, the assured must not come to any arrangement which may prejudice the insurer's rights of subrogation with the third party against whom he has a claim. The insurer's right of subrogation arises not before the insurer has fully indemnified the assured. If any settlement between the assured and the third party is delayed, then the right of subrogation can be delayed as well. Meanwhile, any limitation upon the assured's action equally binds the insurer who has employed the subrogation right.[145]

The doctrine of subrogation is applicable to all contracts of indemnity. Thus, "the doctrine is as applicable to the contracts of indemnity between P&I Clubs and their members as to any other type of contract of insurance."[146] As shipowners' liability insurer, the Club is entitled to be subrogated to the member's right to claim against the third party after the Club has paid compensation to its member in practice.[147]

Therefore, theoretically, in an oil-spill incident, once the liability insurer or other person providing financial security has paid the third parties, he is entitled to obtain all the rights and remedies that the insured has. However, if this is done through subrogation, the effect is limited. In English law, for instance, the insurer is only subrogated to all rights of the assured in respect of the subject-matter insured.[148] Suppose, for example, that ship A came into collision with ship B, for

---

143 See Section B.I of this chapter.

144 Relevant national insurance law or marine insurance law provides specific legislation dealing with subrogation, for instance, in section 79(1) of the England Marine Insurance Act 1906 provides: "(1) Where the insurer pays for a total loss either of the whole, or in the case of goods of any apportionable part, of the subject-matter insured, he thereupon becomes entitled to take over the interest of the insured in whatever may remain of the subject-matter so paid for, and he is thereby subrogated to all the rights and remedies of the assured in and in respect of that subject-matter as from the time of the casualty causing the loss. (2) Subject to the foregoing provisions, where the insurer pays for a partial loss, he acquires no title to the subject-matter insured, or such part of it as may retain, but he is thereupon subrogated to all rights and remedies of the assured in and in respect of the subject-matter insured as from the time of the casualty causing the loss, in so far as the assured has been indemnified, according to this Act, by such payment for the loss."

145 Bennett, Howard, *supra*, note 7, p. 409.

146 Hazelwood, Steven J., *supra*, note 13, p. 303.

147 For more details regarding the Clubs' rights of subrogation, see Hazelwood, Steven J., *ibid.*, in Chapter 12: "The Club's Rights of Subrogation", pp. 303-308.

148 Lambeth, R J, *supra*, note 11, p. 458; Arnould, Joseph/Mustill, Michael J., *supra*, note 22, p. 1089: "Thus, where a vessel is damaged by collision, and her owners

which collision ship A was solely to blame. The incident also caused bunker-oil pollution. B had arranged liability insurance for pollution damage under the Bunkers Convention. Subsequently B has received from the owner of A a sum of money as compensation for the damage to his vessel. However, according to the above-mentioned limit of subrogation, this recovery should not go to that insurer for pollution damage even if he has paid compensation directly to the pollution victims, since for B, the compensation recovered from A was not for pollution damage.

In addition, it is important to avoid shipowners' unjust enrichment. It may be possible for the shipowner to have double recovery both from the third party who was wholly liable for pollution damage and from the liability insurer. This is a mistaken overpayment. Therefore, the insurers shall subrogate the shipowner's right of recourse in this case once they have paid the compensation.

According to the general doctrine of insurance law, the assured must not prejudice the insurer's right of subrogation. There is no express provision in the Bunkers Convention to oblige the assured to take steps to protect the insurer's interest such as preserving time limits and so on. The P&I Clubs, however, have taken precautionary measures in this regard. The current position is that, on the occurrence of a casualty which is likely to give rise to a claim for which a member would seek indemnity from the Club, a member should protect the Club's right of subrogation. Even if there is no express rule, the member should act as a "prudent uninsured shipowner."[149]

G. Concluding remarks

The right to sue directly against the insurer is very important for pollution victims. Nevertheless, eligible claimants should be aware of the time limit for the right to claim compensation. The right of the third party to claim against the insurer shall be extinguished unless an action is brought thereunder within three years from the date when the damage occurred.[150] However, in no case shall an action be brought more than six years from the date of the incident which caused the damage. Where the incident consists of a series of occurrences, the six-year period shall run from the date of the first such occurrence.[151] The insurer will not be liable for any claim which is already time-barred.[152]

recover from those by whose negligence the collision was caused damages in respect of matters which are not covered by a policy on hip, the underwriters cannot, by paying for a total loss, recover from their assured sums paid to them by the wrongdoer, but not paid as part of the value of the ship insured."

149 Hazelwood, Steven J., *supra*, note 13, p. 306.

150 The Bunkers Convention, Art.8.

151 The Bunkers Convention, Art.8.

152 The Bunkers Convention, Art. 8.

Chapter 7: Outlook on Insurance and Compensation for Bunker-Oil Pollution Liability

A. Introduction

In general, all ships carrying bunkers must have the potential to cause pollution damage. It is true that if a shipowner has potentially enormous liability, which would be even disastrous to his business, he would insure himself against such a risk. The insurer, on the other hand, can be held liable only for the amount corresponding to the risks and on the terms that are applicable for his business.

Under the Bunkers Convention, the strict liability imposed on the shipowner is futile unless the insurer agrees to insure the liability,[1] except for the case where the shipowner has the capacity to be a self-insurer. The registered owner shall take out insurance and provide evidence that he has done so. The insurance represents the only guaranteed source of compensation for innocent pollution victims.[2] Besides, the right of direct action conferred on third parties is not in accordance with the practice of marine liability insurance. It intends mainly to protect the interests of innocent victims. However, the exercise of the right of direct action is strictly limited.[3] In addition, the limitation amount of the shipowner's liability is related to the limited insurance capacity of the insurance market.[4]

---

1 The shipowner is required to take out insurance according to the Convention. However, if the ship has a gross tonnage of less than 1,000, he is not required to take out insurance; or even if the ship has a gross tonnage greater than 1,000 but does not belong to a State Party, he is not required to take out insurance either. See Chapter 4.

2 Two points need to be repeated here: first, the registered owner is entitled to maintain insurance or other financial security: Art. 7(1); other financial securities are not discussed in my thesis. Secondly, although the existence of a "joint and several liability" provision in the Convention allows the victims to claim against other jointly liable parties, the compensation from other parties is not necessarily insured, i.e. not secured.

3 See Chapter 6.

4 See Chapter 5.

B. Main interests in the insurance system

Apparently, pollution victims of any bunker-oil spill incident will be in a better position once the Bunkers Convention enters into force. However, the interests of the pollution liability insurers cannot be neglected. It is generally held that the establishment or development of any liability insurance and compensation system in relation to the operation of the ships shall take into account and balance all the interests involved, namely, the pollution victims, the insurer and the shipowner.

I. Victims: "Loss of cover" and full compensation

1. Loss of cover

If the shipowner has taken out insurance, "loss of cover" is a main concern for victims. Pollution victims are much more likely to be distressed by the loss of shipowner's liability cover than by the loss of any other type of insurance cover, since it is the former that deals with related liability coverage for them. Any actual loss of insurance cover obviously presents a danger to innocent victims in an oil-pollution incident.

There are many circumstances under which the assured will lose cover or the insurance contract can cease to provide cover in respect of future occurrence of the kind of risks that would have been covered by the policy. For instance, according to English marine insurance law, a contracting party is entitled to decline to perform his obligations where the other party has failed to comply with a condition precedent to his liability or has substantially failed to perform an intermediate term of the contract.[5] "Loss of cover" can also happen when the insurance coverage provided by the insurance policy is terminated due to the expiration of the period for which the parties agreed. When insurance cover ceases due to reasons such as those described above in an oil spill incident, all innocent pollution victims may find themselves without redress if the shipowner has no assets to answer for his liability. This may usually be illustrated in the case of a one-ship company, which, after the incident, has no other remaining assets except the wreck of the ship.

With the CLCs setting up the background of good examples and experience in compensating for pollution damage within their scope, the Bunkers Convention is also designed to ensure the payment of adequate, prompt and effective compensation for the damage caused by pollution resulting from the escape or discharge of bunker oil from ships. Many aspects in this regard are detailed in the provisions of the Convention. Therefore, once the Bunkers Convention comes into force, the "loss of cover" may not be a concern for most victims, since it will only be in some extreme cases that the shipowner will be deprived of the insurance cover. Nevertheless, the importance of the financial standing of providers of

5 Rose, F. D., *Marine Insurance: Law and Practice* (2004), p. 223.

insurance or financial security is further observed.[6] Caution is also shown to ensure "the completeness and accuracy of the insurance certificate so issued."[7] A certificate attests that insurance or other financial security is in force according to the Bunkers Convention. The period of validity of the certificate shall not be longer than the period of validity of the insurance or other security.[8] For instance, Article 7(3)(c) of the Convention provides, *inter alia*, that:

> "The institution or organisation authorised to issue certificates in accordance with this paragraph shall, as a minimum, be authorised to withdraw these certificates if the conditions under which they have been issued are not maintained. In all cases the institution or organisation shall report such withdrawal to the State on whose behalf the certificate was issued."[9]

2. *"Small ship" issue*

The Convention only obliges the registered owner of the ship having a gross tonnage greater than 1,000 to maintain insurance or other similar financial security. In other words, the ships of 1,000 gross tonnage or less are not obliged to purchase insurance or institute other types of financial security. However, the bunker oil used in small ships may also cause substantial pollution damage. All victims in those cases will run the danger of not being paid compensation.

The reason why small ships are excluded from the obligation to take out insurance is mainly to alleviate the administrative burden which would have been involved in the compulsory insurance regime.[10] The Bunkers Convention, however, does not expressly prohibit any national legislation from imposing such an obligation on the registered owner of small ships. However, the enactment of more national legislation in this respect will ultimately affect the purpose of the Bunkers Convention to have a uniform set of rules.

3. *Adequate compensation*

At present, it is not easy to define "adequate compensation" partly because few disastrous bunker-oil spill incidents have ever occurred. The Bunkers Convention intends to establish a system which can provide adequate compensation for victims who are likely to be adversely affected by serious incidents of a bunker-oil spill. The desire to have adequate compensation would have prompted further considerations to ensure compensation from other jointly liable persons; or to suggest other sources of compensation from the persons or industries involved, or even the States.[11]

---

6 The Bunkers Convention, Art.7(8) and Art.7(9).
7 The Bunkers Convention, Art.7(3)(a).
8 The Bunkers Convention, Art.7(2)(f).
9 The Bunkers Convention, Art.7(3)(c).
10 See Chapter 2, Section B.II.
11 See Chapter 4.

According to the Bunkers Convention, there is no second-tier compensation source along with compensation paid by the compulsory insurance; the insurance will thus become the most frequently used source of compensation for victims. A "joint and several liability" provision of the Convention provides the possibility of other sources of compensation. The availability, however, has the risk of uncertainty. The eligible claimants have no guarantee that they will successfully obtain compensation from persons other than the registered owner. Therefore, the compensation amount presented by the Bunkers Convention is uncertain and its adequacy is also questionable.

In order to protect themselves, the victims may consider taking out insurance themselves. They may choose to insure the loss or damage which can be caused by a bunker-oil spill. This is called first-party insurance, where the insurer's duty is to indemnify the insured for his direct loss. Its purpose is to pay the assured – in this case, victims – for his actual loss and restore him to the state as if the loss had not occurred. This can place the victim in the position previous to the loss, but not in a better position. Basically, when considering purchasing insurance, the assured can choose the insurer and the types of risks he wants to be covered. This is, however, easier said than done, since there are a number of obstacles. One obstacle is the identification of the assured in this case. The assured in the proposed first-party insurance scheme must be the person who will possibly suffer pollution damage as a result of a bunker spill incident. The Bunkers Convention does not, however, have a definition of "pollution victim". An oil-pollution incident may cause pollution damage to the fishermen, hotel owners and so on, who are mostly likely unaware of pollution damage until the incident occurred.[12] Even if those potential pollution victims could have been aware of this situation, the efficiency of such a system would still depend on the availability of the insurer who is willing to insure.

Nevertheless, the above-suggested first-party insurance, if it were to work, could not replace the compulsory insurance under the Bunkers Convention. The better means for all concerned would be to preserve the shipowner's compulsory insurance; and pollution victims may choose to insure their own risks. This would accordingly give both the shipowner and pollution victims the desired certainty of compensation. Meanwhile, in any case, the insurers in the first-party insurance and the compulsory insurance could claim against each other: for instance, if the damage was caused by the fault of the shipowner, the insurer in the first-part insurance could claim compensation from the insurer of the compulsory insurance.

It remains to be seen whether the compensation amount under the Bunkers Convention is adequate in reality. The development of the CLCs shows that the compensation amount for tanker-oil pollution damage has also been amended a number of times in order to increase and satisfy the adequate amount of compensation. However, it cannot be denied that the victims' position will certainly be improved once the Bunkers Convention enters into force.

[12] See Chapter 3, Section C.II.1.

II. The P&I Clubs: maintaining their sustainable development

As a general matter, the Clubs derive their strength from the effective management of the shipowners resorting to the Club's cover. P&I Clubs undertake the insurance obligation for most ships under international civil liability conventions. They will possibly assume a much heavier workload after the ratification of the Bunkers Convention. The Clubs even once asserted that: "...There is the real possibility of creating a large, expensive and cumbersome bureaucracy to deal with a minimum problem, which is likely to persist even after the bureaucracy is in place."[13]

In accordance with international civil liability conventions, the Clubs shall assist their members to obtain insurance certificates as required under the conventions. Besides, the Clubs need to establish a large network of claim-handling agents or correspondents, while more specific technical expertise to assess the risk involved will also be required. However, it is also observed that the P&I system is strengthened rather than weakened by the increase in liability. The reason is that liability conventions let the interests of the insurance practice prevail.[14] This may be true to some extent. For instance, although P&I Clubs have to face victims' claims directly, the victims' right to claim directly is limited. Any action must be related to pollution damage as defined in the Bunkers Convention and the Club will not accept direct action above the insurance limit.[15]

Once the Bunkers Convention comes into force, compulsory insurance will be an indispensable part and different types of ships will face the same obligation to take out insurance. Therefore, the stable and sustainable development of P&I insurance is of significant importance.

III. Shipowners: the central actor

The shipowner plays a central role in liability insurance for bunker-oil pollution damage. He is the one who purchases insurance from his P&I Club, in which he acts as the insurer and assured at the same time. Meanwhile, the insurance purchased is to ensure compensation for innocent pollution victims. Although the liability related to compensation for pollution victims can be insured, it is at a price. The shipowner has to pay his Club more money as an "overspill call" if

13 See IMO LEG 76/3.

14 Røsæg, Erik, 'The Impact of Insurance Practices on Liability Conventions', *Scandinavia Institute of Maritime Law*, originally published in: Legislative approaches in maritime law. Proceedings from the European Colloquium on Maritime Law: Lysebu, Oslo, 7-8 December 2000. MarIus No 283, available at: <http://folk.uio.no/erikro/WWW/corrgr/insurance/Lysebu.pdf> (visited 13 June 2005).

15 However, regarding the limitation of liability under the Bunkers Convention, the Club must pay attention to applicable international or national rules. It can be up to a very high limit. Luckily, the insurer's exposure is in any case not to exceed the specified amount provided for in the 1976 LLMC, as amended. See the Bunkers Convention, Article 7(1).

there was a catastrophic incident in a policy year. Therefore, the ultimate way to protect the shipowner himself is to reduce pollution incidents.

Many measures have been undertaken internally and externally by the Club to reduce pollution risks. The Clubs of the International Group have taken measures in order to maintain the standards of ships and thus reduce potential incidents. For example, the modern system of individual underwriting of the Club takes great account of the individual characteristics and requirements of the entrant shipowner so that one group of members is not subsidised by another.[16] Meanwhile, P&I Clubs also conduct surveys and inspections to ensure that the existing shipowner members continue to maintain the required standards.[17] There is also a detailed legal framework which relates to pollution prevention on a national and international level. The shipowner shall obey the rules therein. An example is the International Convention on Oil Pollution Preparedness, Response and Co-operation (OPRC), 1990.[18] However, they are beyond the topic of this research and hence will not be discussed. The shipowner is also under pressure due to the widely recognised "polluter pays" principle, which has been discussed in Chapter 4.[19] It was agreed that, as a type of financial incentive, it might be a more dynamic way of obtaining improvements within the overall environmental performance of the maritime industry.[20] However, it was concluded that the shipowner under the Bunkers Convention should not be regarded as the "polluter." It is also important to keep in mind that the purpose of the Bunkers Convention is to facilitate compensation being paid for the victims' suffering in an efficient manner. It is not designed as a means of penalising the shipowner.

Ships must continue to improve their standards for environmental protection in order to maintain a competitive position, but there is still a possibility that some ships are insured by the weaker insurers and continue to trade to the disadvantage of other shipowners of good quality. Although there are many external motivations, the measures taken by shipowners to minimise pollution should rather be voluntary and on their own initiative.

C. The comparison of the insurer's exposure to the CLCs and the Convention

This section intends to give a brief comparison of the insurer's liability exposure under the CLCs and under the Bunkers Convention. Most provisions regarding the

---

16 Hazelwood, Steven J., *P&I Clubs Law and Practice* (2000), pp. 112-114.

17 See Chapter 4, Section D.III.2.b).

18 It was adopted by the IMO in November 1990 and entered into force in May 1995. The full text is available in: *Lloyd's Shipping Law Library: the Ratification of Maritime Conventions* (2004), I.7.230.

19 See Chapter 4, Section D.I.

20 'Shipping must Adapt to Environmental Challenge', *Lloyd's List*, September 17 1996.

requirement of compulsory insurance or financial security are similar in both convention systems. The difference will be shown in the following list:

I. The types of ships involved

The CLC Protocol 1992 covers any seagoing vessel constructed or adapted for the carriage of oil in bulk as cargo, i.e. oil tankers. It applies not only during the ship's carriage of oil in bulk, but also during any voyage following such carriage with cargo debris.[21] Therefore, the owners of oil tankers are required to take out insurance. Under the Bunkers Convention, by contrast, the liability insurer will insure any type of seagoing vessel and seaborne craft, except for oil tankers.[22]

II. The number of ships involved

The exposure of the liability insurer due to the ratification of the Bunkers Convention will be much greater than that under the CLCs, since a large number of ships will be covered. The number of ships to be involved largely depends on the insurance threshold. The higher the insurance threshold, the lower the number of ships will be. Although the insurance threshold in the Bunkers Convention is much higher than some States expected, the number of ships involved will still be huge. This will also increase the relevant administrative burden involved in insuring these ships.

III. The limitation of liability

The limitation rule should be as clear and certain as possible. This aspect is made quite clear for tanker owners as well as their insurers. The tanker owner's liability is limited to a fixed amount under the CLCs. The insurer's right to limit his liability is to the same extent. By contrast, the Bunkers Convention enables the insurer to limit his liability in accordance with the provisions of any applicable national or international regime. This may bring uncertainty. Meanwhile, the shipowner's exposure to the liability resulting from a bunker-oil spill does not necessarily denote his insurer's exposure to the same. The guarantee laid down in Articles 7(1) and 7(10) is that even if the shipowner is not entitled to limitation of liability, the insurer is eligible to limit his liability to an amount calculated under the 1976 LLMC, as amended.

To sum up, the insurer's exposure regarding the types of ship involved, the number of ships involved and the right to limit liability is more extensive in the forthcoming Bunkers Convention than it is under the CLCs. The insurer needs to

---

21 The CLC Protocol 1992, Art. 2(1).

22 The Bunkers Convention, Art. 1(1).

pay pollution victims directly. This burden imposed on the liability insurers is exacerbated since they will be the frequently sued targets for compensation.

D. Concluding remarks

The research shows different interests that are involved in an instance of bunker-oil spill, among which are the interests of the victim, the insurer and the ship-owner. A convention cannot come into force unless it gives consideration to relevant interests. There are also other interests from a much higher level, such as the interests of individual States. For instance, the States with vulnerable coastal lines would possibly wish to apply the Bunkers Convention as strictly as possible, but this may not be the case for land-locked States. Furthermore, even States in a similar situation may have different attitudes regarding the ratification of the Convention. An international convention often cannot satisfy all States in the world. Nevertheless, the adoption or ratification of a convention needs to satisfy at least the interests of most of the States. An interesting example, which is indicative of this level of satisfaction, relates to the negotiation on the threshold of insurance. The insurance threshold, as we understand now, was a compromise and it failed to satisfy some countries. The arguments which resulted in such a compromise became somewhat three-dimensional when it was linked to the questions of how many ratifying States should be needed before the Bunkers Convention comes into force and whether the number of States specified for entry into force should possess a certain minimum tonnage of registered vessels in order to trigger the entry into force requirements.[23]

As mentioned, there are so far 9 Contracting States to the Bunkers Convention representing 9.07% of world tonnage.[24] Compared with the requirement of 18 States to be for the ratification of the Convention, there is still a long way to go.[25] However, it is important to recall that the Civil Liability and the Fund Conventions covering liability and compensation for oil spills from tankers took six and seven years respectively before coming into effect, although they required only eight ratifying States in order for them to enter into force.

In September 2002, the Council of the EU Commission made a decision to authorise the Member States, in the interest of the Community, to sign, ratify or accede to the Bunkers Convention.[26] The authorisation was considered necessary,

---

23 Patrick Griggs, 'International Convention on Civil Liability for Bunker Oil Pollution Damage', available at: <http://www.bmla.org.uk/documents/imo-bunker-convention.htm> (visited 15 June 2005).

24 <http://www.imo.org/Conventions/mainframe.asp?topic_id=247> (visited 7 January 2006). These 9 Contracting States include: Cyprus, Greece, Jamaica, Latvia, Luxemburg, Samoa, Slovenia, Spain, and Tonga. See Chapter 1, Section C.VII.

25 The Bunkers Convention, Art.14(1).

26 See Council Decision of 19 September 2002, 2002/762/EC, available at: <http://europa.eu.int/eur-lex/lex/staging/LexUriServ/LexUriServ.do?uri=CELEX:32002D0762:EN:HTML> (visited 22 December 2005).

since there was a legislative gap. The EU Council Regulation No 44/2001 awards the EU exclusive responsibility for all matters pertaining to jurisdiction and the recognition and enforcement of judgments in civil and commercial matters and is binding for the Member States.[27] As a consequence, the Member States do not have the authority to ratify the Bunkers Convention by themselves as only the Community has the right to negotiate, conclude and fulfil the international commitments related to matters of jurisdiction in civil and commercial matters. However, only States, as sovereign States, can become contracting parties to the Bunkers Convention. Therefore a situation would arise whereby neither any EU Member State nor the Community would be in a position to approve the Convention. The Council decision especially authorising ratification of the Bunkers Convention by the Member States was thus proposed in order to solve the problem. Pursuant to the said Council decision, Member States shall make efforts to sign the Bunkers Convention before 30 September 2002, and they shall take the necessary steps to deposit the instruments of ratification of, or accession to, the Bunkers Convention within a reasonable time with the Secretary-General of the IMO and, if possible, before 30 June 2006.[28] It was a signal of support from one part of the international community for the ratification of the Bunkers Convention.

The goals pursued by the Bunkers Convention cannot be attained unless the Bunkers Convention is ratified and comes into force. The purpose of the Bunkers Convention is to provide uniformity instead of maintaining or developing the situation of having different national legislation in handling liability issues and providing adequate compensation for bunker-oil pollution damage. The effect of the Bunkers Convention, once it comes into force, will depend on the efficiency of relevant provisions in the Convention and their implementation, in particular that of compulsory insurance.

---

27 Council Regulation (EC) No 44/2001 of 22 December 2000, the document is available at: <http://europa.eu.int/eur-lex/pri/en/oj/dat/2001/l_012/l_01220010116en00010023.pdf> (visited 22 December 2005).

28 See Article 3 in the Council Decision of 19 September 2002 (2002/762/EC).

Summary of Study

Blame for oil pollution mostly focuses on oil tankers. The system of liability and compensation in the Civil Liability Convention (CLC) and Fund Convention, established after the *Torrey Canyon* incident in 1967, is the most widespread regime for addressing the issue of oil-pollution liability and compensation. International harmonisation of civil liability for oil pollution is of significant importance. It ensures that common liability standards apply in all the contracting States. Consequently, it can contribute to simplifying the burden facing pollution victims, while at the same time clarifying the liabilities of shipowners. It also creates an equitable balance between different industry interests.

However, oil tankers should not be the only vessels to be blamed for causing oil pollution at sea. Some spills that caused expensive pollution damage in the past have been of heavy fuel oil from non-tankers. There was thus a need to bring the law on marine oil pollution up-to-date by extending liability and compensation to all seagoing vessels. In March 2001, the International Maritime Organization adopted a new convention which was entitled "International Convention on Civil Liability for Bunker Oil Pollution Damage, 2001" (the Bunkers Convention). The Bunkers Convention is largely modelled on the earlier international civil liability conventions. Under the Bunkers Convention, owners of different types of ships other than oil tankers are strictly liable for pollution damage caused by bunker oil spilled from their ships. Shipowners have the right to limit their liability. At the same time, compulsory insurance for oil-pollution liability is imposed on the registered owner. The Bunkers Convention has not yet come into force: it will come into force after eighteen States, including five States each with ships not less than 1 million gross tonnage, have ratified it.[1] It is hard to estimate the effect of the Bunkers Convention at this stage, but it is safe to point out that the adoption of the Bunkers Convention has satisfied the need to address liability and compensation issues arising from the spills of bunker oil from ships and, once ratified, it will fill the liability gap left by the earlier civil liability conventions.

The aim of this thesis is to examine compulsory insurance and compensation for pollution damage under the Bunkers Convention. It is necessary to indicate that the analysis mainly focuses on compulsory insurance and compensation issues, so other issues, that are likewise important under the Bunkers Convention, are not considered in this thesis. Since the Bunkers Convention has already been

1 The Bunkers Convention, Art. 14(1).

adopted pending ratification, any criticisms or suggestions in this thesis are only for purposes of academic research.[2]

In order to achieve the stated aim, the thesis is divided into seven chapters. Chapter 1 presents an overview of the Bunkers Convention after a brief description of previous industry solutions and international conventions. Chapter 2 deals with the birth of compulsory insurance and discusses the insurer for oil-pollution liability. The Bunkers Convention recognises the need for compulsory insurance and requires the registered owner to do so. The shipowners' P&I Clubs, the main liability insurers for shipowners, will be expected to insure liability for pollution damage under the Bunkers Convention. Their insurance practice in cases of tanker oil spills is described in order to show how it would be possible for the Clubs to handle future bunker-oil pollution claims. At the same time, the external challenge from other insurance providers due to the enactment of the Oil Pollution Act of 1990 (OPA 90) in the U.S. is also discussed.[3] Chapter 3 aims to analyse the interrelations between strict liability and insurance. It also explores the reasons for applying the strict liability rule for oil pollution and the liability allocations under the Bunkers Convention. Chapter 4 provides an in-depth discussion of insurance and the quest for adequate compensation. Apart from the discussion of the administrative burden corollary to the issuance of the insurance certificate, the lack of a second-tier compensation source and the uncertain amount of compensation under the Bunkers Convention lead to the question of whether the Bunkers Convention will offer sufficient compensation for victims. Other possible sources of compensation are discussed in this chapter, but they are only for purposes of academic discussion, since they did not appear with the adoption of the Bunkers Convention. Chapter 5 deals with limitation of liability and the limit of insurance. The shipowner's right to limit his liability is affirmed due to the insurability consideration. This chapter, however, scrutinises the possibility of modifying the limitation rule of the Bunkers Convention and is aimed at considering whether it would be possible to make available a higher amount of compensation. This, however, is only for purposes of academic research, too. Although the victims' right to claim directly against the liability insurer is an indispensable element of the compulsory insurance, Chapter 6 provides a separate examination of the right of direct action against the insurer and its limited effect. The comparison is made between the general right of a third party to claim on the insurance policy, direct action under P&I insurance and the direct-action right under the Bunkers Convention. Besides, the concept of "pollution damage" is also examined. Chapter 7 provides an outlook on insurance and compensation. This chapter concludes the thesis by pointing out that it is important to strike a balance of interests among the pollution victims, the shipowner and the liability insurer.

---

2 Tsimplis, Michael N., 'The Bunker Pollution Convention 2001: Completing and Harmonizing the Liability Regime for Oil Pollution from Ships', *Lloyd's Mar. & Com. L.Q.* (2005), pp. 83-100, at 99: "It is likely that many of these issues will have to be clarified in the future if they cause difficulties in practice."

3 The U.S. do not accept the international civil liability system.

Through these seven chapters, there are two main points: first, the concept of compulsory insurance and its compensation purpose; second, other features of the Bunkers Convention and their interrelations with compulsory insurance.

A. The concept of compulsory insurance and its compensation purpose

Before 1969, insurance for any liability was an internal matter for shipowners.[4] After long and protracted discussions, the shipowner's liability insurance for oil-pollution damage was made compulsory with the advent of the 1969 CLC. Accordingly, the owner of the ship having a gross tonnage above a certain threshold shall take out insurance or have other financial security to adequately meet his potential pollution liability under victims' claims. This concept is followed by the Bunkers Convention. It provides in Article 7 of the Convention that the registered owner of a ship, which is defined as "any seagoing vessel and seaborne craft, of any type whatsoever"[5], having a gross tonnage greater than 1,000 and registered in a State Party shall be required to maintain insurance or other financial security.[6] Besides liability insurance, the shipowner can also maintain other financial security to cover his liability. Other financial security may be maintained by different methods. There are no specific provisions regarding a uniform interpretation of what constitutes the acceptable providers of financial security, but one method deemed sufficient is a guarantee from a bank.[7] More details about other types of financial security are not discussed in this thesis. The insurance or other financial security has to be certified. The certificate attests that insurance or other financial security is in force in accordance with the provisions of the Bunkers Convention.[8] There was a worry that the Bunkers Convention, once ratified, would involve a much larger number of ships of different types, thus intensifying the administrative burden. The implied administrative burden was, however, deemed to be minimal as compared to the advantages it will bring to the State Parties, such as having guaranteed compensation available once the incident occurs in their port area.

As an integral part of the compulsory insurance mechanism, claims for compensation are entitled to be brought directly by third parties against the insurers or other parties named in the certificate as the guarantor.[9] The direct-action right under the Bunkers Convention is more than a right to enforce a claim directly as a way of avoiding procedural complexity and expenses. It is mainly to

---

4 Rosag, Erik, 'Compulsory Maritime Insurance', *Scandinavian Institute of Maritime Law Yearbook 2000*, available at: <http://folk.uio.no/erikro/WWW/corrgr/insurance/simply.pdf> (visited 24 May 2005), at 1.

5 The Bunkers Convention, Art.1(1).

6 The Bunkers Convention, Art.7(1).

7 The Bunkers Convention, Art.7(1).

8 The Bunkers Convention, Art.7(2).

9 The Bunkers Convention, Art.7(10).

ensure that prompt and effective compensation is available to pollution victims. The insurer or guarantor has a very limited scope of defence. However, eligible claimants are liable to be defeated by any defence invoked by the insurer which the shipowner would have been entitled to invoke, save only that the Bunkers Convention nullifies clauses which purport to relieve the insurer from liability upon the shipowner's insolvency. The exposure of the insurer or guarantor will not exceed the liability limit applying to the ship. In addition, the insurer or guarantor may invoke the defence that pollution damage resulted from the wilful misconduct of the shipowner.[10] In the absence of a definition in the Bunkers Convention, the meaning of the term "wilful misconduct" may depend on which national court decides on the issue.

The shipowners' P&I Club provides insurance services for shipowners' third-party liabilities and expenses arising from the owning and operation of their ships. It is a mutual and non-profit association. The P&I Club insures shipowners' pollution liability, including oil pollution; it also provides the necessary evidence that the insurance cover for oil-pollution liability is in place as required, which enables the certificate to be issued by relevant government authorities in accordance with the CLCs. In practice, the duty of the P&I Club to indemnify routinely follows the "pay to be paid rule", i.e. the assured is not entitled to be paid by the Club until the assured himself has made payment to the third parties. However, the right of direct action conferred on third parties under the CLCs and the Bunkers Convention is not in accordance with the usual practice of the P&I Club, since this right mainly serves the purpose of protecting the interests of pollution victims. It has, nevertheless, been observed in the thesis that the effect of the right of direct action is limited. In particular, it is important to emphasise the limitation of liability terms in the pollution liability clause. This is due to the apprehension of the Club that possible disastrous pollution damage may be caused in an oil-spill incident to the detriment of its business. Nevertheless, the limit offered by the Club has been raised from US$500 million to US$ 1000 (US$1 billion) for each occurrence for each vessel in the case of oil pollution in 2000. On the worldwide market, shipowners' P&I Clubs are not the only providers of liability insurance for shipowners. Other liability insurance providers have emerged on the U.S. market due to the rigorous financial responsibility requirement in the OPA 90.

The liability insurance is usually taken out by the shipowner to protect himself against oil-pollution damage to victims. It is, however, noticed that the most important purpose of compulsory insurance under the Bunkers Convention is to ensure that compensation is available and sufficient for pollution victims. It is difficult to define adequate compensation. The compensation has to be sufficient for a particular case. It follows from the provisions in the Bunkers Convention that compensation may be from different parties, such as the bareboat charterer, operator and manager,[11] but the liability insurance taken out by the registered owner represents the only guaranteed source of compensation for pollution victims. Apart from this, no other supplementary or complementary guaranteed

[10] The Bunkers Convention, Art.7(10).
[11] The Bunkers Convention, Art.3(1) and Art. 3(2).

source of compensation appeared with the adoption of the Bunkers Convention or at a later stage. Therefore, the Bunkers Convention pays particular attention to ensure the financial standing of providers of insurance or financial security for the purpose of the Bunkers Convention.[12] In addition, compensation is paid only for pollution damage which is defined in Article 1(9) of the Bunkers Convention. Due to the similar definition in the earlier CLCs, the interpretation of "pollution damage" in the Bunkers Convention will make reference to the CLCs. The fundamental difference is that the oil which causes pollution damage must be bunker oil, which is defined as "any hydrocarbon mineral oil, including lubricating oil, used or intended to be used for the operation or propulsion of the ship, and any residues of such oil."[13]

B. Other features of the Bunkers Convention and their interrelations with compulsory insurance

The main features of the Bunkers Convention, like other civil liability conventions, are strict liability, compulsory insurance and limitation of liability. They operate together under the Bunkers Convention.

It is strict liability that the insurance covers. The choice of strict liability by the Bunkers Convention was made without much debate as in earlier civil liability conventions, since it is an almost invariable feature of all international liability conventions. The principle of strict liability means that the shipowner must be liable for pollution damage irrespective of any fault on his part. It was observed that the person who could most easily be identified should be the one to be held liable and to take out insurance. This idea is applied in the CLCs. However, the distribution of liabilities and the insurance obligation are different under the Bunkers Convention. It requires the shipowner, a term which includes the bareboat charterer, manager and operator, to be liable for pollution damage,[14] while imposing insurance obligation only on the registered owner.[15] This seems to be a difficult decision. Arguably, the bareboat charterer may be able to take out his own insurance, it is, however, not practical to require the manager and operator to take out insurance, since the identities are frequently changed during transit. Such provisions can, however, increase the possibility for victims to claim compensation from them, since they are required to be liable jointly and severally.[16]

Limitation of liability is a traditional principle in maritime law. One of the main reasons to follow this principle was that insuring the shipowner's liability risk is greatly facilitated when the maximum liability can normally be assessed in advance.[17] Although the capping of the insurer's exposure can provide a con-

12 The Bunkers Convention, Art.7(8).
13 The Bunkers Convention, Art.1(5).
14 The Bunkers Convention, Art.3(1) and Art.1(3).
15 The Bunkers Convention, Art.7(1).
16 The Bunkers Convention, Art.3(2).
17 Documentation C.M.I. 1968-III, at 146.

siderable degree of comfort to the insurance industry and also speed up the settlement of claims, the limitation of the owner's liability should not necessarily be tied up with the insurance obligation; therefore, it is suggested in the thesis that the insurance coverage could be limited while the whole liability is left unlimited. Nevertheless, the idea of "unlimited liability" was of no interest to the drafters of a bunkers convention. In addition, it should be highlighted that the Bunkers Convention does not have its own limitation regime and thus the amount of limitation is virtually uncertain; it will depend on "any applicable national or international regime" in a particular case, which may result in unlimited liability, if applicable.[18] The resolution, which was adopted in this respect with the adoption of the Bunkers Convention, urges all States to ratify the Protocol of 1996 to amend the Convention on Limitation of Liability for Maritime Claims, 1976 (LLMC Protocol 1996). However, the resolution is only recommendatory in nature. The relevant provisions of LLMC, if applicable, would in effect also impact the applicability of the Bunkers Convention.[19]

It is clear that the adoption of the Bunkers Convention indeed ensures that all other types of ships will face similar international rules and procedures as oil tankers for determining questions of liability and providing compensation in oil-spill incidents. Nevertheless, the objective pursued by the Bunkers Convention cannot be attained unless the Bunkers Convention first comes into force. In the meantime, compulsory insurance is a very important and indispensable component and it will mainly ensure that the compensation is available to pollution victims.

18 The Bunkers Convention, Art.6.

19 For instance, in Tsimplis, Michael N., *supra*, note 2, at.99: "Whether LLMC 1976, Art 2(1) covers damage from bunker oil pollution when no physical damage has been sustained also remains unclear and this impacts on the applicability of BOPC." BOPC denotes the Bunkers Convention.

Bibliography

Literature

Abcassis, David W. (ed.), *Oil Pollution from Ships* (2nd ed., London: Stevens & Sons, 1985)

Abraham, Kenneth S., 'Environmental Pollution Liability Insurance Scenario: Toward Predictability', in: *The Role of Insurance and Compensation in Environmental Pollution Problems: Workshop Papers May 15, 1986*, Centre for Research on Risk and Insurance, the Wharton School, University of Pennsylvania with the contributions of the Geneva Association, pp. 53-55

Abraham, Kenneth S., 'Environmental Liability and the Limits of Insurance', *Colum. L. Rev:* (1988), pp. 942-988

Alcantara, Leonard F./Cox, Mary A., 'OPA 90 Certificates of Financial Responsibility', *23 J. Mar. L. & Com.* 369 (1992), pp. 369-386

Anderson, Charles B./DelaRue, Colin "Liability of Charterers and Cargo Owners for Pollution from Ships", 26 *Tul. Mar. L. J.* (2001), pp. 1-60

Ansell, D.V./Dicks, B., *et al.* (ITOPF), 'A Review of the Problems Posed by Spills of Heavy Fuel Oil', available at: <http://www.itopf.com/iosc2001.pdf> (visited 31 March 2005)

Arnould, Joseph/Mustill, Michael J., *Arnould's Law of Marine Insurance and Average*, Volume 1 (16th ed., London: Sweet & Maxwell, 1981)

Arnould, Joseph/Mustill, Michael J., *Arnould's Law of Marine Insurance and Average*, Volume 2 (16th ed., London: Sweet & Maxwell, 1981)

Athearn, James L., *Risk and Insurance* (St. Paul, etc.: West Publishing Company, 1977)

Baughen, Simon, *Shipping Law* (3rd ed., London: Cavendish, 2004)

Beddard, Ralph, 'The Implementation of the Convention', in: *The Limitation of Shipowner's Liability: the New Law* (London: Sweet & Maxwell, 1986), pp. 152-160

Bennett, Howard N., *The Law of Marine Insurance* (Oxford: Clarendon Press, 1996)

Bennett, Paul, 'Mutual Risk: P&I Insurance Clubs and Maritime Safety and Environmental Performance', *Marine Policy* 25 (2001), pp. 13-21

Bergkamp, Lucas, *Liability and Environment: Private and Public Law Aspects of Civil Liability for Environmental Harm in an International Context* (The Hague: Kluwer Law International, 2001)

Blanpain, R. (General ed.), *International Encyclopaedia of Laws: Insurance law* (Deventer: Kluwer Law and Taxation Publ., 1992)

Bocken, H., 'Developments with respect to Compensation for Damage Caused by Pollution', in: Markesinis, Basil S. (ed.), *The Gradual Convergence: Foreign Ideas, Foreign Influences, and English Law, on the Eve of the 21st Century* (Oxford: Clarendon Press, 1994)

Braekhus, sjur and Rein, Alexander, *Handbook of P&I Insurance* (2nd ed., Arendal: Assuranceforeningen Gard-Gjensidig, 1979)

Brodecki, Z., 'New Definition of Pollution Damage', *Lloyd's Mar. & Com. L. Q.* (1985), pp. 382-391

Brown, Robert H., *Introduction to Marine Insurance: Training Notes for Brokers* (2nd ed., London: Witherbys Publishing, 1995)

Bryant, Hugh, 'Specialist Insurers Offer Real OPA Solution for Shipowners', *Lloyd's List*, Friday, June 21 1996

Chen, Xia, *Limitation of Liability for Maritime Claims: a Study of U.S. Law, Chinese Law, and International Conventions* (The Hague: Kluwer Academic Publishers, 2001)

Clark, Robert B, *Marine Pollution* (5th ed., Oxford: Oxford University Press, 2003)

Clark, A.F. Bessemer , 'The U.S. Oil Pollution Act of 1990', *Lloyd's Mar. & Com. L.Q.* (1991), pp. 247-256

Clarke, Malcolm A., *The Law of Insurance Contracts* (4th Ed, London: LLP, 2002)

Coles, Richard M. F./Ready, Nigel P. (eds.), *Ship Registration: Law and Practice* (London: LLP, 2002)

Coghlin, T. G., 'Protection & Indemnity Clubs', *Lloyd's Mar. & Com. L.Q.* (1984), pp. 403-416

Colton, N.J., 'The Underwriting of Oil Pollution Risks', in DeLaRue, Colin M. (ed.), *Liability for Damage to the Marine Environment* (London: Lloyd's of London Press, 1993), pp. 149-154

Davis, Mark, *Bareboat Charters* (London: LLP, 2000)

DelaRue, Colin M. (ed.), *Liability for Damage to the Marine Environment* (Lloyd's London Press, 1993)

DelaRue, Colin M./Anderson, Charles B., *Shipping and the Environment: Law and Practice* (London: LLP, 1998).

Derrington, D.K, *The Law of Liability Insurance* (Sydney: Butterworth, 1990)

Dougherty, Daniel J., 'The Impact of a Member's Insolvency or Bankruptcy on a Protection & Indemnity Club', in: *59 Tul. L. Rev. 1466*, pp. 1466-1486

Dykes, Andrew, 'Limitation and Oil Pollution', in: *The Limitation of Shipowner's Liability: the New Law* (London: Sweet & Maxwell, 1986), pp. 144-151

Faure, Michael, 'Economic Analysis', in: Koch, Bernhard A./Koziol (eds.), *Unification of Tort Law: Strict Liability* (The Hague: Kluwer Law International, 2002)

Faure, Michael (ed.), *Deterrence, Insurability, and Compensation in Environmental Liability: Future Developments in the European Union* (Wien: Springer, 2003)

Fedtke, Joerg/Magnus, Ulrich, 'Germany', in: Faure, Michael, 'Economic Analysis', in: Koch, Bernhard A./Koziol (eds.), *Unification of Tort Law: Strict Liability* (The Hague: Kluwer Law International, 2002), pp. 147-176

Forster, Malcolm, 'Civil Liability of Shipowners for Oil Pollution', *J.B.L.*1973, pp. 23-31

Fossion, Gregory, 'An Eternal Triangle at Sea: Loss of Insurance Cover under a Direct Action in Marine Liability Insurance', available at: <http://www.law.kuleuven.ac.be/jura/39n2/fossion.htm> (visited 10 October 2005)

Francioni, Francesco & Tullio Scovazzi (eds.), *International Responsibility for Environmental Harm* (London: Graham & Trotman, 1991)

Galand-Carval, Suzanne, 'France', in: Koch, Bernhard A./Koziol (eds.), *Unification of Tort Law: Strict Liability* (The Hague: Kluwer Law International, 2002), pp. 127-145

Gauci, Gotthard, *Oil Pollution at Sea: Civil Liability and Compensation for Damage* (Chichester: Wiley, 1997)

Gauci, Gotthard, 'Limitation of Liability in Maritime Law: an Anachronism?' *Maritime Policy*, Vol.19, No.1, pp. 65-74

Geoffrey Brice, Q.C., 'The Scope of the Limitation Action', in: *The Limitation of Shipowners' Liability: the New Law* (London: Sweet & Maxwell, 1986), pp. 18-32

Gold, Edgar, *Gard Handbook on P&I Insurance* (5th ed., Arendal: Gard Assuranceforeningen, 2002)

Gold, Edgar/Chircop, Aldo/Kindred, Hugh, *Canadian Maritime Law* (Toronto: Irwin Law, 2003)

Griggs, Patrick/Williams, Richard, *Limitation of Liability for Maritime Claims* (3rd ed., London: Lloyd's of London Press, 1998)

Goldhirsch, Lawrence B., *The Warsaw Convention Annotated: a Legal Handbook* (2nd ed., The Hague: Kluwer Law International, 2000)

'Leading article: Good Oil', *Lloyd's* list, October 14 1998

Griggs, Patrick, 'Limitation of Liability for Maritime Claims: the Search for International Uniformity', *Lloyd's Mar. & Com. L.Q.* (1997), pp. 369-377

Griggs, Patrick, 'International Convention on Civil Liability for Bunker Oil Pollution Damage', available at: <http://www.bmla.org.uk/documents/imo-bunker-convention.htm> (visited 15 June 2005)

Grime, Robert, 'Implementation of the 1976 Limitation Convention: Liability for Maritime Claims', *Marine Policy*, July 1988, pp. 306-313

Grime, Robert: 'The loss of the Right to Limit', in: *The Limitation of Shipowners' Liability: the New Law* (London: Sweet & Maxwell, 1986), pp. 102-112

Gyselen, Luc, 'P&I insurance: the European Commission's Decision Concerning the Agreements of the International Group of P&I Clubs', Head of the Unite in charge of 'Financial Services', Directorate-general for Competition, European Commission. In: *Marc Huybrechts, Eric Van Hooydonk and Christian Dieryck (eds.), Marine Insurance at the turn of the Millenium*, Volume 1 (Antwerpen: Intersentia, 1999), pp. 181-202

Hazelwood, Steven J., *P&I Clubs: Law and Practice* (3rd ed, London: Lloyd's of London Press, 2000)

Hill, Christopher, *Maritime Law* (6th ed., London: LLP, 2003)

Hill, Christopher/Robertson, Bill/Hazelwood, Steven J., *Introduction to P&I* (2nd ed., London: Lloyd's of London Press, 1996)

Hodges, Susan, *Cases and Materials on Marine Insurance Law* (London: Cavendish Publishing, 2004)

Holmes, Eric M. (general ed.), *Holmes' Appleman on Insurance*, Volume Fifteen, Chapters 111.1-115.3 (2nd ed., St. Paul, etc.: West, 2000)

Hudson, N. Geoffrey (ed.), *The Institute Clauses* (London: LLP, 1999)

Hurd, Howard B, *The Law and Practice of Marine Insurance relating to Collision Damages and Other Liabilities to Third Parties* (2nd ed., London: Pitman, 1952)

Hutchinson, James A., 'Financial Responsibility Provisions: are They Sinking the U.S. Maritime Trade?' *24 Law & Pol'y Int'L Bus.* 223

Ivamy, E.R.Hardy, *General Principles of Insurance Law* (6th Ed., London: Butterworths, 1993)

Jaap Molenaar, Erik, *Coastal State Jurisdiction over Vessel-Source Pollution* (The Hague: Kluwer Academic Publishers Group, 1998)

Jackson, David, 'The 1976 Convention and International Uniformity of Rules', in: *The New Law: Limitation of Shipowners' Liability* ((London: Sweet & Maxwell, 1986), pp. 126-143

Johnson, H. Alston, 'The Louisiana Direct Action Statute', 43 *La.L.Rev.*1455

Keeton, G.W., 'The Lessons of the Torrey Canyon: English Law Aspects', in: *Current Legal Problems 1968*, Volume 21, pp. 94-112

Kierr, Raymond H, 'The effect of Direct Action Statutes on P&I Insurance, on Various other Insurances of Maritime Liabilities, and on Limitation on Shipowners' Liability', 43 *Tul. L. Rev.*, pp. 638-672

Kim, Inho, 'Financial Responsibility Rules under the Oil Pollution Act of 1990', 42 *Nat. Resources J.* 565

Knight, F., Risk, *Uncertainty and Profit* (New York: Houghton Mifflin, 1921)

Koch, Bernhard A./Koziol (eds.), *Unification of Tort Law: Strict Liability* (The Hague: Kluwer Law International, 2002)

Kunreuther, Howard, 'The Role of Insurance and Compensation in Environmental Pollution Problems', workshop papers, May 15, 1986, Center for Research on Risk and Insurance, the Wharton School, University of Pennsylvania with the contributions of the Geneva Association, pp. 1-6

Landes, W.M. and Posner, R. A., *The Economic Structure of Tort Law* (Cambridge, Mass. Etc.: Harvard University Press, 1987)

Larsson, Maria-Louise, *The Law of Environmental Damage: Liability and Reparation* (Edsbruk: Stockholm University Dissertation, 1997)

Li, KX & Ingram, CWM, *Maritime Law and Policy in China* (London: Cavendish, 2002)

Lockett, Nick, *Environmental Liability Insurance* (London: May, 1996)

Maginnis, Michael J./Cot, Jose R., 'Direct Action Statutes and P&I Insurance', available at: <http://www.mlaus.org/article.ihtml?id=576&folder=90> (visited on 22 February 2006)

Mandaraka-Sheppard, Alexandra, *Modern Admiralty Law* (London: Cavendish, 2001)

Mankiewicz, Rene H, *The Liability Regime of the International Air Carrier* (Deventer, Frankfurt. etc.: Kluwer, 1981)

Markesinis, B.S.(ed.), *The Gradual Convergence: Foreign Ideas, Foreign Influences, and English Law, on the Eve of the 21st Century* (Oxford: Clarendon Press, 1994)

Merkin, Robert M. (ed.), *Colinvaux's Law of Insurance* (7th ed., London: Sweet & Maxwell, 1997)

Mitchell, John M., 'The United States Coastal Guard's Proposed Regulation of Certificates of Financial Responsibility under the Oil Pollution Act of 1990: Fostering a Continuing Market of Insurance for shipowners?' 7 *Admin. L. J. Am. U.*121

Mo, John Shijian, *Shipping Law in China* (Hong Kong: Sweet & Maxwell Asia, 1999)

Molenaar, Erik Jaap, *Coastal State Jurisdiction over Vessel-Source Pollution* (The Hague: Kluwer Law International, 1998)

Morrison, Fred L./Wolfrum, Rüdiger (eds.), *International, Regional and National Environmental Law* (The Hague: Kluwer Law International, 2000)

Moss, Gabriel (ed.), *Cross-frontier Insolvency of Insurance Companies* (London: Sweet & Maxwell, 2001)

'Oil Pollution a Dead Issue?' 2 *Marine Pollution Bulletin*, Issue 9, September 1971, pp. 129-130

O'May, Donald, *Marine Insurance: Law and Policy* (London: Sweet & Maxwell, 1993)

Özçayir, Z. Oya, *Liability for Oil Pollution and Collisions* (London: LLP, 1998)

Özçayir, Z. Oya, *Port State Control* (London: LLP, 2001)

Özçayir, Z. Oya: 'Limitation of Liability Problems in cases of Oil Pollution', available at: <http://www.turkishpilots.org/DOCUMENTS/Oya_Ozcayir_Liability.htm> (visited 8 June 2005)

Perron, Edgar du/ Boom, Willem H.van, 'Netherlands', in: Koch, Bernhard A./Koziol (eds.), *Unification of Tort Law: Strict Liability* (The Hague: Kluwer Law International, 2002), pp. 227-255

Pfennigstorf, W., 'Policy Considerations for Insurers Engaging in Environmental Liability Insurance', in: H. Bocken/D. Ryckbost (Eds.), *Insurance of Environmental Damage* (Brussel: Story-Scientia, 1991), pp. 269-289

Pfennigstorf, W., 'Limited Insurability of Unlimited Liability: Serial Claims, Aggregates and Alternatives: the Continental View', in: Ralph P Kroener (ed.), *Transnational Environmental Liability and Insurance* (London: Graham& Trotman, 1993), pp. 59-165

Pfennigstorf, Werner/Gifford, Donald G., *A Comparative Study of Liability Law and Compensation Schemes in Ten Countries and the United States* (Oak Brook, III: Insurance Research Council, 1991)

Prosser, William Lloyd/Keeton, W. Page, *Prosser and Keeton on the Law of Torts* (5th ed., St Paul: West, 1984)

UK P&I Club, 'Port State Control', *P&I Int.*, 2003, 17(5), 14

Readman, Luke, 'Certificate of Financial Responsibility under OPA 90: the Catalyst for an Insurance Crisis', *Int. I. L. R.*1994, 2(10), 361-364

Richardson, Benjamin J., 'Mandating Environmental Liability Insurance', 12 *Duke Envtl. L. & Pol'y F.* 293, at 296

Rogers, W.V.H., *Winfield and Jolowicz on Tort* (14th Edition, London: Sweet & Maxwell, 1994)

Rogers, W.V.H., 'England', in: B.A. Koch and H.Koziol (eds.), *Unification of Tort Law: Strict Liability* (The Hague: Kluwer Law International, 2002)

Rose, F. D., *Marine Insurance: Law and Practice* (London: LLP, 2004)

Røsæg, Erik, 'Compulsory Maritime Insurance', *Scandinavian Institute of Maritime Law Yearbook 2000*, available at: <http://folk.uio.no/erikro/WWW/corrgr/insurance/simply.pdf> (visited 24 May 2005)

Røsæg, Erik, 'The Impact of Insurance Practices on Liability Conventions', *Scandinavia Institute of Maritime Law*, originally published in: Legislative approaches in maritime law. Proceedings from the European Colloquium on Maritime Law: Lysebu, Oslo, 7-8 December 2000. MarIus No 283, available at: <http://folk.uio.no/erikro/ WWW/corrgr/insurance/Lysebu.pdf> (visited 13 June 2005)

Sands, Philippe J., *Principles of International Environmental Law I: Frameworks, Standards and Implementation* (Manchester: Manchester University Press, 1995)

Schattke, H., 'The Legal Aspects of the Equitable Distribution of Limited Indemnification Amounts', *Nuclear Third Party Liability and Insurance, Status and Prospect*, Munich Symposium, Paris: OECD, 1965, Vol. 97

Schoenbaum, Thomas J., *Key divergences between English and American law of Marine Insurance* (Centreville, Md: Cornell Maritime Press, 1999)

Selvig, Erling, 'An Introduction to the 1976 Convention', in: *The Limitation of Shipowners' Liability: the New Law* (London: Sweet & Maxwell, 1986), pp. 3-17

Seward, R.C., 'The Insurance Viewpoint', in: *The Limitation of Shipowners' Liability: the new law* (London: Sweet & Maxwell, 1986), pp. 161-186

Seward, R. C., 'The Role of Protection and Indemnity (P&I) Clubs', available at: <http://www.intertanko.com/pubupload/protection%20%20indemnity%20HK%202002.pdf#search='the%20role%20of%20P&I%20club> (visited 14 April 2005)

'Shipping Must Adapt to Environmental Challenge', *Lloyd's List*, September 17, 1996

Speares, Sandra, 'Change to 1930 Act Offer Hope for Claims against Insolvent Insurers', *Lloyd's List*, September 18 2002

Springer, Allen L., *The International Law of Pollution: Protecting the Global Environment in a World of Sovereign States* (Westport, Conn. etc., Quorum Books, 1983)

Spurrier, Andrew, 'Call for Reform of Liability Insurance', *Lloyd's List*, Monday January 08 2001

Steel, David, 'Ships are Different: the Case for Limitation of Liability', *Lloyd's Mar. & Com. L.Q.* (1995), pp. 77-87

Tanega, Joseph, 'Implications of Environmental Liability on the Insurance Industry', 8 *J. Envtl. L.*115, pp. 115-137

Templeman, Frederick, *Templeman on Marine Insurance: Its Principles and Practice* (6th ed., London: Pitman, 1986)

Thomas, David Rhidian (ed.), *The Modern Law of Marine Insurance* (London: LLP, 2002)

Tsimplis, Michael N., 'The Bunker Pollution Convention 2001: Completing and Harmonizing the Liability Regime for Oil Pollution from Ships?' *Lloyd's Mar. & Com. L.Q.* (2005), pp. 83-100

Vandall, Frank J., *Strict Liability: Legal and Economic Analysis* (New York etc.: Quorum Books, 1989)

Vicuna, Francisco Orrego, 'Responsibility and Liability for Environmental Damage under International Law: Issues and Trends', 10 *Geo. Int'l Envtl. L. Rev.*, pp. 279-308

Werro, Franz/Palmer, Vernon Valentine (eds.), *The Boundaries of Strict Liability in European Tort Law* (Bern: Staempfli, 2004)

Winfield, Percy Henry, *Winfield and Jolowicz on Tort* (12th ed., London: Sweet & Maxwell, 1984)

Willingale, Malcolm, *Ship Management* (London: LLP, 1998)

Wolfrum, Ruediger/Roeben, Volker/Morrison, Fred L., 'Preservation of the Marine Environment', in: Morrison, Fred L. and Wolfrum, Ruediger (eds.): *International, Regional and National Environmental Law* (The Hague: Kluwer Law International, 2000).

Wagner, Thomas J., 'The Oil Pollution Act of 1990: an Analysis', 21 *J. Mar. L. & Com.* 569 (1990), pp. 569-587

Wetterstein, Peter, 'P&I and Environmental Damage', in: *The Seventh Axel Ax: son Johnson Colloquium on Maritime Law Hässelby Colloquium, May, 27-28, 1993: P&I Insurance*, pp. 115-163

White, Ian C., 'Factors Affecting the Cost of Oil Spills', available at <http://www.itopf.com/costs02.PDF> (visited 31 March 2005)

Wilde, Mark, *Civil Liability for Environmental Damage: a Comparative Analysis of Law and Policy in Europe and the United States* (The Hague: Kluwer, 2002)

Wu, Chao, *Pollution from the Carriage of Oil by Sea: Liability and Compensation* (London: Kluwer Law International, 1996)

Wu, Chao, 'Liability and Compensation for Bunker Pollution', 33 *J. Mar. L. & Com.* (2002), pp. 553-567

Documents

1. IMO documents

IMCO official records of the international legal conference on marine pollution damage, 1969.
Official Records of the International Conference on the Limitation of Liability for Maritime Claims, 1976.
IMO Ref. T3/2.01: Revised list of certificates and documents required to be carried on board ships.
IMO Publication: Manual on Oil Pollution: Section IV-Combating Oil Spills.

IMO LEG 73/12/1
IMO LEG 74/4/1
IMO LEG 74/4/2
IMO LEG 74/4/3
IMO LEG 74/4/4
IMO LEG 75/5/1
IMO LEG 76/WP.3 Agenda item 4
IMO LEG 75/5/11
IMO LEG 75/WP.1
IMO LEG 77/4/3
IMO LEG 77/4/5
IMO LEG 77/4/6
IMO LEG 77/4/7
IMO LEG 77/6
IMO LEG 77/6/1
IMO LEG 77/6/2
IMO LEG 77/11
IMO LEG 77/WP.3 (Agenda 6)
IMO LEG 78/5/1
IMO LEG 78/5/2
IMO LEG 78/5/3
IMO LEG 78/5/4
IMO LEG 78/11
IMO LEG 78/WP.3 (Agenda 5)
IMO LEG 79/4/2
IMO LEG 79/6
IMO LEG 79/6/1
IMO LEG 79/6/2
IMO LEG 79/6/3
IMO LEG 79/11
IMO LEG 80/4
IMO LEG 80/4/1
IMO LEG 80/4/2
IMO LEG 80/10/5
IMO LEG 81/4
IMO LEG 81/4/1
IMO LEG 81/4/2
IMO LEG 81/4/3
IMO LEG 81/4/4
IMO LEG 81/4/5
IMO LEG 81/10
IMO LEG 81/WP.2 (Agenda 4)

IMO LEG 82/3
IMO LEG 82/3/1
IMO LEG 82/3/2
IMO LEG 82/3/3
IMO LEG 82/10/3
IMO LEG 83/11
IMO LEG 83/13/2
IMO LEG/CONF. 12/4
IMO LEG/CONF. 12/6
IMO LEG/CONF. 12/7
IMO LEG/CONF. 12/8
IMO LEG/CONF. 12/9
IMO LEG/CONF. 12/10
IMO LEG/CONF. 12/12
IMO LEG/CONF. 12/13
IMO LEG/CONF. 12/14
IMO LEG/CONF. 12/15
IMO LEG/CONF. 12/16
IMO LEG/CONF. 12/17
IMO LEG/CONF. 12/18
IMO LEG/CONF. 12/19/Corr.1
IMO LEG/CONF. 12/20
IMO LEG/CONF. 12/NF.2.

2. CMI documents

C.M.I., Documentation 1968.I.
C.M.I., Documentation 1968. III.
CMI Guidelines on Oil Pollution Damage.

3. IOPC Fund documents

IOPC Fund, general information on liability and compensation for oil pollution damage, February 1992.
IOPC Fund Annual Report 1997.
IOPC Fund Annual Report 1998.
IOPC Fund, Claims Manual, November 2002.
IOPC Fund's Claims Manual (2002), available at: <http://www.iopcfund.org/npdf/92claim.pdf> (visited 12 June 2005).

4. P&I Club documents

UK P&I Club: Analysis of Major Claims 1993.

The Rule Book of North of England, 2004/2005, available at: <http://195.173.17.24/pdf/PandI_Rules_2005.pdf> (visited 25 May 2005).

The UK Club's Rules 2005, available at: <http://www.ukpandi.com/ukpandi/resource.nsf/Files/Rules05_02/$FILE/Rules05_02.pdf> (visited 26 May 2005).

5. Other documents

OECD document, Maritime Transport Committee report on the Removal of Insurance from substandard Shipping, June 2004.

Commission of the European Communities, Communication from the Commission to the Council and Parliament and the Economic and Social Committee: Green Paper on Remedying Environmental Damage, Brussels, 14 May 1993, COM (93) 47 final.

Institute Cargo Clauses (ICC).

The Institute Time Clauses (Hulls).

Minster of Land, Infrastructure and Transport of Japan, 'Introduction of Compulsory Insurance Requirement for Non-tanker Ships', available at: <http://www.mlit.go.jp/english/maritime/Web/insurance_portal3eng.htm> (visited 17 November 2005).

The UK Department of Transpor, 'Consultation on Implementation of the Bunkers Convention', available at: <http://www.dft.gov.uk/stellent/groups/ dft_control/documents/contentservertemplate/dft_index.hcst?n=14405&l=2> (visited on 24 December 2005).

The European Union, Council Decision of 19 September 2002, 2002/762/EC, see at: <http://europa.eu.int /eur-lex /lex/staging /LexUri Serv/LexUriServ.do?uri=CELEX:32002D0762:EN:HTML> (visited 22 December 2005).

The European Union, Council Regulation (EC) No. 44/2001 of 22 December 2000, available at: <http://europa.eu.int/eur-lex/pri/en/ oj/dat/2001/l_012/l_01220010116en00010023.pdf> (visited 22 December 2005).

Appendix: Text of the Bunkers Convention

International Convention on Civil Liability for Bunker Oil Pollution Damage, 2001

The States Parties to this Convention,

Recalling article 194 of the United Nations Convention on the Law of the Sea, 1982, which provides that States shall take all measures necessary to prevent, reduce and control pollution of the marine environment,

Recalling also article 235 of that Convention, which provides that, with the objective of assuring prompt and adequate compensation in respect of all damage caused by pollution of the marine environment, States shall co-operate in the further development of relevant rules of international law,

Noting the success of the International Convention on Civil Liability for Oil Pollution Damage, 1992 and the International Convention on the Establishment of an International Fund for Compensation for Oil Pollution Damage, 1992 in ensuring that compensation is available to persons who suffer damage caused by pollution resulting from the escape or discharge of oil carried in bulk at sea by ships,

Noting also the adoption of the International Convention on Liability and Compensation for Damage in Connection with the Carriage of Hazardous and Noxious Substances by Sea, 1996 in order to provide adequate, prompt and effective compensation for damage caused by incidents in connection with the carriage by sea of hazardous and noxious substances,

Recognising the importance of establishing strict liability for all forms of oil pollution which is linked to an appropriate limitation of the level of that liability,

Considering that complementary measures are necessary to ensure the payment of adequate, prompt and effective compensation for damage caused by pollution resulting from the escape or discharge of bunker oil from ships,

Desiring to adopt uniform international rules and procedures for determining questions of liability and providing adequate compensation in such cases,

Have agreed as follows:

Article 1
Definitions

For the purposes of this Convention:

1 "Ship" means any seagoing vessel and seaborne craft, of any type whatsoever.

2 "Person" means any individual or partnership or any public or private body, whether corporate or not, including a State or any of its constituent subdivisions.
3 "Shipowner" means the owner, including the registered owner, bareboat charterer, manager and operator of the ship.
4 "Registered owner" means the person or persons registered as the owner of the ship or, in the absence of registration, the person or persons owning the ship. However, in the case of a ship owned by a State and operated by a company which in that State is registered as the ship's operator, "registered owner" shall mean such company.
5 "Bunker oil" means any hydrocarbon mineral oil, including lubricating oil, used or intended to be used for the operation or propulsion of the ship, and any residues of such oil.
6 "Civil Liability Convention" means the International Convention on Civil Liability for Oil Pollution Damage, 1992, as amended.
7 "Preventive measures" means any reasonable measures taken by any person after an incident has occurred to prevent or minimize pollution damage.
8 "Incident" means any occurrence or series of occurrences having the same origin, which causes pollution damage or creates a grave and imminent threat of causing such damage.
9 "Pollution damage" means:
(a) loss or damage caused outside the ship by contamination resulting from the escape or discharge of bunker oil from the ship, wherever such escape or discharge may occur, provided that compensation for impairment of the environment other than loss of profit from such impairment shall be limited to costs of reasonable measures of reinstatement actually undertaken or to be undertaken; and
(b) the costs of preventive measures and further loss or damage caused by preventive measures.
10 "State of the ship's registry" means, in relation to a registered ship, the State of registration of the ship and, in relation to an unregistered ship, the State whose flag the ship is entitled to fly.
11 "Gross tonnage" means gross tonnage calculated in accordance with the tonnage measurement regulations contained in Annex 1 of the International Convention on Tonnage Measurement of Ships, 1969.
12 "Organization" means the International Maritime Organization.
13 "Secretary-General" means the Secretary-General of the Organization.

Article 2
Scope of application

This Convention shall apply exclusively:
(a) to pollution damage caused:
in the territory, including the territorial sea, of a State Party, and
in the exclusive economic zone of a State Party, established in accordance with international law, or, if a State Party has not established such a zone, in an area beyond and adjacent to the territorial sea of that State determined by that State in

accordance with international law and extending not more than 200 nautical miles from the baselines from which the breadth of its territorial sea is measured;
(b) to preventive measures, wherever taken, to prevent or minimize such damage.

Article 3
Liability of the shipowner

1 Except as provided in paragraphs 3 and 4, the shipowner at the time of an incident shall be liable for pollution damage caused by any bunker oil on board or originating from the ship, provided that, if an incident consists of a series of occurrences having the same origin, the liability shall attach to the shipowner at the time of the first of such occurrences.
2 Where more than one person is liable in accordance with paragraph 1, their liability shall be joint and several.
3 No liability for pollution damage shall attach to the shipowner if the shipowner proves that:
(a) the damage resulted from an act of war, hostilities, civil war, insurrection or a natural phenomenon of an exceptional, inevitable and irresistible character; or
(b) the damage was wholly caused by an act or omission done with the intent to cause damage by a third party; or
(c) the damage was wholly caused by the negligence or other wrongful act of any Government or other authority responsible for the maintenance of lights or other navigational aids in the exercise of that function.
4 If the shipowner proves that the pollution damage resulted wholly or partially either from an act or omission done with intent to cause damage by the person who suffered the damage or from the negligence of that person, the shipowner may be exonerated wholly or partially from liability to such person.
5 No claim for compensation for pollution damage shall be made against the shipowner otherwise than in accordance with this Convention.
6 Nothing in this Convention shall prejudice any right of recourse of the shipowner which exists independently of this Convention.

Article 4
Exclusions

1 This Convention shall not apply to pollution damage as defined in the Civil Liability Convention, whether or not compensation is payable in respect of it under that Convention.
2 Except as provided in paragraph 3, the provisions of this Convention shall not apply to warships, naval auxiliary or other ships owned or operated by a State and used, for the time being, only on Government non-commercial service.
3 A State Party may decide to apply this Convention to its warships or other ships described in paragraph 2, in which case it shall notify the Secretary-General thereof specifying the terms and conditions of such application.

4 With respect to ships owned by a State Party and used for commercial purposes, each State shall be subject to suit in the jurisdictions set forth in article 9 and shall waive all defences based on its status as a sovereign State.

Article 5
Incidents involving two or more ships

When an incident involving two or more ships occurs and pollution damage results there from, the shipowners of all the ships concerned, unless exonerated under article 3, shall be jointly and severally liable for all such damage which is not reasonably separable.

Article 6
Limitation of liability

Nothing in this Convention shall affect the right of the shipowner and the person or persons providing insurance or other financial security to limit liability under any applicable national or international regime, such as the Convention on Limitation of Liability for Maritime Claims, 1976, as amended.

Article 7
Compulsory insurance or financial security

1 The registered owner of a ship having a gross tonnage greater than 1000 registered in a State Party shall be required to maintain insurance or other financial security, such as the guarantee of a bank or similar financial institution, to cover the liability of the registered owner for pollution damage in an amount equal to the limits of liability under the applicable national or international limitation regime, but in all cases, not exceeding an amount calculated in accordance with the Convention on Limitation of Liability for Maritime Claims, 1976, as amended.
2 A certificate attesting that insurance or other financial security is in force in accordance with the provisions of this Convention shall be issued to each ship after the appropriate authority of a State Party has determined that the requirements of paragraph 1 have been complied with. With respect to a ship registered in a State Party such certificate shall be issued or certified by the appropriate authority of the State of the ship's registry; with respect to a ship not registered in a State Party it may be issued or certified by the appropriate authority of any State Party. This certificate shall be in the form of the model set out in the annex to this Convention and shall contain the following particulars:
(a) name of ship, distinctive number or letters and port of registry;
(b) name and principal place of business of the registered owner;
(c) IMO ship identification number;
(d) type and duration of security;

(e) name and principal place of business of insurer or other person giving security and, where appropriate, place of business where the insurance or security is established;
(f) period of validity of the certificate which shall not be longer than the period of validity of the insurance or other security.
3 (a) A State Party may authorize either an institution or an organization recognized by it to issue the certificate referred to in paragraph 2. Such institution or organization shall inform that State of the issue of each certificate. In all cases, the State Party shall fully guarantee the completeness and accuracy of the certificate so issued and shall undertake to ensure the necessary arrangements to satisfy this obligation.
(b) A State Party shall notify the Secretary-General of:
(i) the specific responsibilities and conditions of the authority delegated to an institution or organization recognised by it;
(ii) the withdrawal of such authority; and
(iii) the date from which such authority or withdrawal of such authority takes effect.
An authority delegated shall not take effect prior to three months from the date on which notification to that effect was given to the Secretary-General.
(c) The institution or organization authorized to issue certificates in accordance with this paragraph shall, as a minimum, be authorized to withdraw these certificates if the conditions under which they have been issued are not maintained. In all cases the institution or organization shall report such withdrawal to the State on whose behalf the certificate was issued.
4 The certificate shall be in the official language or languages of the issuing State. If the language used is not English, French or Spanish, the text shall include a translation into one of these languages and, where the State so decides, the official language of the State may be omitted.
5 The certificate shall be carried on board the ship and a copy shall be deposited with the authorities who keep the record of the ship's registry or, if the ship is not registered in a State Party, with the authorities issuing or certifying the certificate.
6 An insurance or other financial security shall not satisfy the requirements of this article if it can cease, for reasons other than the expiry of the period of validity of the insurance or security specified in the certificate under paragraph 2 of this article, before three months have elapsed from the date on which notice of its termination is given to the authorities referred to in paragraph 5 of this article, unless the certificate has been surrendered to these authorities or a new certificate has been issued within the said period. The foregoing provisions shall similarly apply to any modification which results in the insurance or security no longer satisfying the requirements of this article.
7 The State of the ship's registry shall, subject to the provisions of this article, determine the conditions of issue and validity of the certificate.
8 Nothing in this Convention shall be construed as preventing a State Party from relying on information obtained from other States or the Organization or other international organisations relating to the financial standing of providers of insurance or financial security for the purposes of this Convention. In such cases,

the State Party relying on such information is not relieved of its responsibility as a State issuing the certificate required by paragraph 2.

9 Certificates issued or certified under the authority of a State Party shall be accepted by other States Parties for the purposes of this Convention and shall be regarded by other States Parties as having the same force as certificates issued or certified by them even if issued or certified in respect of a ship not registered in a State Party. A State Party may at any time request consultation with the issuing or certifying State should it believe that the insurer or guarantor named in the insurance certificate is not financially capable of meeting the obligations imposed by this Convention.

10 Any claim for compensation for pollution damage may be brought directly against the insurer or other person providing financial security for the registered owner's liability for pollution damage. In such a case the defendant may invoke the defences (other than bankruptcy or winding up of the shipowner) which the shipowner would have been entitled to invoke, including limitation pursuant to article 6. Furthermore, even if the shipowner is not entitled to limitation of liability according to article 6, the defendant may limit liability to an amount equal to the amount of the insurance or other financial security required to be maintained in accordance with paragraph 1. Moreover, the defendant may invoke the defence that the pollution damage resulted from the wilful misconduct of the shipowner, but the defendant shall not invoke any other defence which the defendant might have been entitled to invoke in proceedings brought by the shipowner against the defendant. The defendant shall in any event have the right to require the shipowner to be joined in the proceedings.

11 A State Party shall not permit a ship under its flag to which this article applies to operate at any time, unless a certificate has been issued under paragraphs 2 or 14.

12 Subject to the provisions of this article, each State Party shall ensure, under its national law, that insurance or other security, to the extent specified in paragraph 1, is in force in respect of any ship having a gross tonnage greater than 1000, wherever registered, entering or leaving a port in its territory, or arriving at or leaving an offshore facility in its territorial sea.

13 Notwithstanding the provisions of paragraph 5, a State Party may notify the Secretary-General that, for the purposes of paragraph 12, ships are not required to carry on board or to produce the certificate required by paragraph 2, when entering or leaving ports or arriving at or leaving from offshore facilities in its territory, provided that the State Party which issues the certificate required by paragraph 2 has notified the Secretary-General that it maintains records in an electronic format, accessible to all States Parties, attesting the existence of the certificate and enabling States Parties to discharge their obligations under paragraph 12.

14 If insurance or other financial security is not maintained in respect of a ship owned by a State Party, the provisions of this article relating thereto shall not be applicable to such ship, but the ship shall carry a certificate issued by the appropriate authority of the State of the ship's registry stating that the ship is owned by that State and that the ship's liability is covered within the limit

prescribed in accordance with paragraph 1. Such a certificate shall follow as closely as possible the model prescribed by paragraph 2.
15 A State may, at the time of ratification, acceptance, approval of, or accession to this Convention, or at any time thereafter, declare that this article does not apply to ships operating exclusively within the area of that State referred to in article 2(a)(i).

Article 8
Time limits

Rights to compensation under this Convention shall be extinguished unless an action is brought thereunder within three years from the date when the damage occurred. However, in no case shall an action be brought more than six years from the date of the incident which caused the damage. Where the incident consists of a series of occurrences, the six-years' period shall run from the date of the first such occurrence.

Article 9
Jurisdiction

1 Where an incident has caused pollution damage in the territory, including the territorial sea, or in an area referred to in article 2(a)(ii) of one or more States Parties, or preventive measures have been taken to prevent or minimise pollution damage in such territory, including the territorial sea, or in such area, actions for compensation against the shipowner, insurer or other person providing security for the shipowner's liability may be brought only in the courts of any such States Parties.
2 Reasonable notice of any action taken under paragraph 1 shall be given to each defendant.
3 Each State Party shall ensure that its courts have jurisdiction to entertain actions for compensation under this Convention.

Article 10
Recognition and enforcement

1 Any judgement given by a Court with jurisdiction in accordance with article 9 which is enforceable in the State of origin where it is no longer subject to ordinary forms of review, shall be recognised in any State Party, except:
(a) where the judgement was obtained by fraud; or
(b) where the defendant was not given reasonable notice and a fair opportunity to present his or her case.
2 A judgement recognised under paragraph 1 shall be enforceable in each State Party as soon as the formalities required in that State have been complied with. The formalities shall not permit the merits of the case to be re-opened.

Article 11
Supersession Clause

This Convention shall supersede any Convention in force or open for signature, ratification or accession at the date on which this Convention is opened for signature, but only to the extent that such Convention would be in conflict with it; however, nothing in this article shall affect the obligations of States Parties to States not party to this Convention arising under such Convention.

Article 12
Signature, ratification, acceptance, approval and accession

1 This Convention shall be open for signature at the Headquarters of the Organization from 1 October 2001 until 30 September 2002 and shall thereafter remain open for accession.
2 States may express their consent to be bound by this Convention by:
(a) signature without reservation as to ratification, acceptance or approval;
(b) signature subject to ratification, acceptance or approval followed by ratification, acceptance or approval; or
(c) accession.
3 Ratification, acceptance, approval or accession shall be effected by the deposit of an instrument to that effect with the Secretary-General.
4 Any instrument of ratification, acceptance, approval or accession deposited after the entry into force of an amendment to this Convention with respect to all existing State Parties, or after the completion of all measures required for the entry into force of the amendment with respect to those State Parties shall be deemed to apply to this Convention as modified by the amendment.

Article 13
States with more than one system of law

1 If a State has two or more territorial units in which different systems of law are applicable in relation to matters dealt with in this Convention, it may at the time of signature, ratification, acceptance, approval or accession declare that this Convention shall extend to all its territorial units or only to one or more of them and may modify this declaration by submitting another declaration at any time.
2 Any such declaration shall be notified to the Secretary-General and shall state expressly the territorial units to which this Convention applies.
3 In relation to a State Party which has made such a declaration:
(a) in the definition of "registered owner" in article 1(4), references to a State shall be construed as references to such a territorial unit;
(b) references to the State of a ship's registry and, in relation to a compulsory insurance certificate, to the issuing or certifying State, shall be construed as referring to the territorial unit respectively in which the ship is registered and which issues or certifies the certificate;

(c) references in this Convention to the requirements of national law shall be construed as references to the requirements of the law of the relevant territorial unit; and
(d) references in articles 9 and 10 to courts, and to judgements which must be recognized in States Parties, shall be construed as references respectively to courts of, and to judgements which must be recognized in, the relevant territorial unit.

Article 14
Entry into Force

1. This Convention shall enter into force one year following the date on which 18 States, including five States each with ships whose combined gross tonnage is not less than 1,000,000, have either signed it without reservation as to ratification, acceptance or approval or have deposited instruments of ratification, acceptance, approval or accession with the Secretary-General.
2 For any State which ratifies, accepts, approves or accedes to it after the conditions in paragraph 1 for entry into force have been met, this Convention shall enter into force three months after the date of deposit by such State of the appropriate instrument.

Article 15
Denunciation

1 This Convention may be denounced by any State Party at any time after the date on which this Convention comes into force for that State.
2 Denunciation shall be effected by the deposit of an instrument with the Secretary-General.
3 A denunciation shall take effect one year, or such longer period as may be specified in the instrument of denunciation, after its deposit with the Secretary-General.

Article 16
Revision or amendment

1 A conference for the purpose of revising or amending this Convention may be convened by the Organization.
2 The Organization shall convene a conference of the States Parties for revising or amending this Convention at the request of not less than one-third of the States Parties.

Article 17
Depositary

1 This Convention shall be deposited with the Secretary-General.
2 The Secretary-General shall:
(a) inform all States which have signed or acceded to this Convention of:

(i) each new signature or deposit of instrument together with the date thereof;
(ii) the date of entry into force of this Convention;
(iii) the deposit of any instrument of denunciation of this Convention together with the date of the deposit and the date on which the denunciation takes effect; and
(iv) other declarations and notifications made under this Convention.
(b) transmit certified true copies of this Convention to all Signatory States and to all States which accede to this Convention.

Article 18
Transmission to United Nations

As soon as this Convention comes into force, the text shall be transmitted by the Secretary-General to the Secretariat of the United Nations for registration and publication in accordance with Article 102 of the Charter of the United Nations.

Article 19
Languages

This Convention is established in a single original in the Arabic, Chinese, English, French, Russian and Spanish languages, each text being equally authentic.
Done at London this twenty-third day of March two thousand and one.
In witness whereof the undersigned being duly authorised by their respective Governments for that purpose have signed this Convention.

***

ANNEX

CERTIFICATE OF INSURANCE OR OTHER FINANCIAL SECURITY IN RESPECT OF CIVIL LIABILITY FOR BUNKER OIL POLLUTION DAMAGE

Issued in accordance with the provisions of article 7 of the International Convention on Civil Liability for Bunker Oil Pollution Damage, 2001

Name of Ship	Distinctive Number or letters	IMO Ship Identification Number	Port of Registry	Name and full address of the principal place of business of the registered owner.

This is to certify that there is in force in respect of the above-named ship a policy of insurance or other financial security satisfying the requirements of article 7 of the International Convention on Civil Liability for Bunker Oil Pollution Damage, 2001.

Type of Security

Duration of Security

Name and address of the insurer(s) and/or guarantor(s)

Name

Address

This certificate is valid until

Issued or certified by the Government of

(Full designation of the State)

OR

The following text should be used when a State Party avails itself of article 7(3)

The present certificate is issued under the authority of the Government of .....(full designation of the State) by......(name of institution or organization)

At On

(Place) (Date)

(Signature and Title of issuing or certifying official)

Explanatory Notes:

1. If desired, the designation of the State may include a reference to the competent public authority of the country where the Certificate is issued.

2. If the total amount of security has been furnished by more than one source, the amount of each of them should be indicated.

3. If security is furnished in several forms, these should be enumerated.

4. The entry "Duration of Security" must stipulate the date on which such security takes effect.

5. The entry "Address" of the insurer(s) and/or guarantor(s) must indicate the principal place of business of the insurer(s) and/or guarantor(s). If appropriate, the place of business where the insurance or other security is established shall be indicated.

Index

About the International Max Planck Research School for Maritime Affairs at the University of Hamburg

The International Max Planck Research School for Maritime Affairs at the University of Hamburg was established by the Max Planck Society for the Advancement of Science, in co-operation with the Max Planck Institute for Foreign Private Law and Private International Law (Hamburg), the Max Planck Institute for Comparative Foreign Public Law and International Law (Heidelberg), the Max Planck Institute for Meteorology (Hamburg) and the University of Hamburg. The School's research is focused on the legal, economic, and geophysical aspects of the use, protection, and organization of the oceans. Its researchers work in the fields of law, economics, and natural sciences. The School provides extensive research capacities as well as its own teaching curriculum. Currently, the School has 12 Directors who determine the general work of the School, act as supervisors for dissertations, elect applicants for the School's PhD-grants, and are the editors of this book series:

Prof. Dr. Dr. h.c. Jürgen Basedow is Director of the Max Planck Institute for Foreign Private Law and Private International Law; *Prof. Dr. Peter Ehlers* is the Director of the German Federal Maritime and Hydrographic Agency; *Prof. Dr. Dr. h.c. Hartmut Graßl* is Director of the Max Planck Institute for Meteorology; *Prof. Dr. Hans-Joachim Koch* is Managing Director of the Seminar of Environmental Law of the Faculty of Law at the University of Hamburg; *Prof. Dr. Rainer Lagoni* is Managing Director of the Institute of Maritime Law and the Law of the Sea at the University of Hamburg; *PD Dr. Gerhard Lammel* is Senior Scientist at the Max Planck Institute for Meteorology; *Prof. Dr. Ulrich Magnus* is Managing Director of the Seminar of Foreign Law and Private International Law at the University of Hamburg; *Prof. Dr. Peter Mankowski* is Director of the Seminar of Foreign and Private International Law at the University of Hamburg; *Prof. Dr. Marian Paschke* is Director of the Institute of Maritime Law and the Law of the Sea at the University of Hamburg; *Prof. Dr. Jürgen Sündermann* is Director at the Center for Marine and Climate Research at the University of Hamburg; *Prof. Dr. Richard Tol* is Director of the Research Unit Sustainability and Global Change at the University of Hamburg; *Prof. Dr. Dr. h.c. Rüdiger Wolfrum* is Director at the Max Planck Institute for Comparative Foreign Public Law and International Law and a judge at the International Tribunal for the Law of the Sea.

At present, *Prof. Dr. Dr. h.c. Jürgen Basedow* and *Prof. Dr. Ulrich Magnus* serve as speakers of the International Max Planck Research School for Maritime Affairs at the University of Hamburg.

Batch number: 10761599

Printed by Printforce, the Netherlands